AF330080

ENGINEERING TOOLS, TECHNIQUES AND TABLES

ADVANCES IN SYSTEMS ENGINEERING RESEARCH

ENGINEERING TOOLS, TECHNIQUES AND TABLES

Additional books in this series can be found on Nova's website under the Series tab.

Additional e-books in this series can be found on Nova's website under the e-book tab.

ADVANCES IN SYSTEMS ENGINEERING RESEARCH

ELENA FERMI
AND
ADAM LAMBERTI
EDITORS

For permission to use material from this book please contact us:
Telephone 631-231-7269; Fax 631-231-8175
Web Site: http://www.novapublishers.com

Library of Congress Cataloging-in-Publication Data

Advances in systems engineering research / editors, Elena Fermi and Adam Lamberti.
    pages cm
Includes index.
ISBN 978-1-62948-310-8 (hardcover)
1. Systems engineering. I. Fermi, Elena. II. Lamberti, Adam.
TA168.A295 2011
320.001'1--dc23
             2013038304

Published by Nova Science Publishers, Inc. † *New York*

Contents

PREFACE

Systems engineering is an interdisciplinary field of engineering that focuses on how to design and manage complex engineering projects over their life cycles. Issues such as reliability, logistics, coordination of different teams, evaluation measurements, and other disciplines become more difficult when dealing with large, complex projects. Systems engineering deals with work-processes, optimization methods, and risk management tools in such projects. In this book, the authors discuss putting systems engineering into practice and the model-based systems engineering paradigm; education of military engineers and managers with a systems perspective at the Air Force Institute of Technology; simulation of the separation of industrially important hydrocarbon mixtures by different distillation techniques using Mathematica©; enterprise transformation and addressing individual perceptual and behavioral biases as scaling fractals to create emergent state changes; and the application and validation of systems engineering methods and techniques.

Chapter 1 – This chapter aims to explore the emerging Model-Based Systems Engineering (MBSE) paradigm as the fundamental way to put Systems Thinking and Systems Engineering (SE) into practice.

To engineer the modern large, complex, interdisciplinary systems-of-systems, the collaborative world teams must "speak" the same language and must work on the same "matter". The "matter" is the System Model and the communication mechanisms must be supported by standard, flexible, and friendly modelling languages. The evolving MBSE approach is leading the way and it is expected to become a standard practice in the Systems Engineering field in the next decade. The future of MBSE will be facilitated by the continuously evolving information technologies (computing power, storage and analysis capacities, distributed capabilities, virtual networking, etc.) as well as by the fine tuned profile of systems engineers (the proliferation of SE courses at the various graduation levels and the adaptive profile innate to the new generations will contribute to the Systems Engineer of the future).

As a recent trend within SE field, MBSE does not have, by this time, a dedicated piece of fundamental theory. Nevertheless, there is a universal idea behind all the ongoing published research efforts: the critical importance of modelling to engineer complex socio-technical systems.

This chapter constitutes an attempt to gather the more relevant science in the topic (e.g., main definitions, standards, visual modelling languages, methodologies) and also presents a new agile methodology (LITHE - Agile Systems Modelling Engineering) to enhance the application of MBSE.

Chapter 2 – The Systems Engineering and Management Department of the Air Force Institute of Technology (AFIT) is advancing systems engineering through the weaving of education, collaborative research, and consultation processes in its Systems Engineering Research and Analysis Group (SERAG). The SERAG integrates faculty and graduate students from three curriculums; Graduate Engineering Management, Graduate Environmental Engineering and Science, and Graduate Systems Engineering, with each having experienced and multi-disciplinary civilian and military faculty, and a variety of laboratory facilities and test environments. The SERAG's synergistic integration is yielding innovative research results providing enhanced national defense capability for platforms and processes as well as military officers and DoD civilians who carry with them increased knowledge and a systems perspective upon their graduation from AFIT. The integrative approach utilized by the SERAG enhances national defense by conceiving and developing new concepts focused on the organization, training, and equipping of the Air Force that will aid the Air Force warfighter of today and of tomorrow. This chapter details the three educational courses of study themselves, summarizes several of the system-level products of technical collaborations and consultations with external clients, and describes results of the most recent collaborative research projects of the SERAG. In addition, this chapter argues for Air Force use of a systems view when identifying assignments for graduating students. Use of the systems view for assignments would align the newly acquired knowledge of the graduating student with the needed skills of the assignment accelerating innovation and organizational transformation.

Chapter 3 – In this chapter the simulation of several case studies for the separation by distillation of various industrially relevant hydrocarbon mixtures are worked out. Different techniques used to separate the complex mixtures are illustrated: classical column train and enhanced distillation (extractive and reactive).

This chapter is aimed primarily at chemical engineering students. But the information presented might be also useful for postgraduate and PhD students as well as professional engineers.

An important objective of this chapter is to show how these rather complex separations can be handled with great pedagogical benefits using the computer algebra Mathematica$^{©}$, a general purpose solver, not chemical engineering specific. Simulation results are compared to those obtained using the flow-sheeting software Aspen-HYSYS$^{®}$.

Five case studies are worked out: the separation of natural gas liquids (NGL), the fractionation of a C_4 cut to separate 1,3-butadiene with furfural as entrainer, the production of MTBE from i-butene and methanol, the reverse process: production of i-butene and methanol by the decomposition of methyl $tert$-butyl ether (MTBE), and the equilibrium-limited metathesis of cis-2-pentene to cis-2-butene and cis-2-hexene.

For every case complete information is given such that it can be easily reproduced with the process simulator and/or alternatively worked out using the relevant Mathematica© programs available at http://demonstrations.wolfram.com.

Chapter 4 – Organizations are complex adaptive systems with basic elements of individual persons, deciding and acting according to their own personal logic and behavior. Effective collective action is known to best occur when individuals are aligned toward a unifying corporate vision. Cognitive biases and heuristics, however, have been evolutionarily embedded into individual decision making so as to minimize individual risks, as exhibited through behavioral focus on short-term goals, bounded rationality, and the use of intuition as acquired through personal experience. The transition from a personal level of perception to a collective behavioral space and the contextual organizational setting is mediated by the Fundamental Attribution Error, the Correspondence Bias, and the Actor-Observer Bias at the cognitive level. The organizational effects of ameliorating and modifying individual biases are predicted through Balance Theory models that account for interpersonal behavioral changes that lead toward cognitive consistency among individuals and that can eventually lead to emergent and lasting enterprise transformations. The principal aim of this research is disrupt the traditional paradigm of dissipative managerial direction with demonstrable alternatives that enable common and beneficent personal decisions to rapidly scale up, creating holistic state changes in enterprises.

Chapter 5 – The Systems Engineering Body of Knowledge (SEBoK) is often described as a set of best practices. However, literature provides little substantiation for applicability and value of the captured methods and techniques. A research question is when these methods and techniques are applicable, and how much they improve performance of project teams. Research challenges are to collect data in the real world -where most influencing factors are not under control of the researchers- and to cope with the heterogeneity of the field of application. For example, the system-of-interest can be an integrated circuit or a space station; it can be well bounded or embedded in a complex socio-technical environment.

This chapter reports the authors' research in an industrial cluster, where the authors have sampled the application of various systems engineering methods and techniques in a variety of domains. The authors discuss the research methodology, and they show a number of emerging trends after taking 67 samples in industrial practice in Kongsberg. Examples of applied methods and techniques are requirements management, interface management, modeling, concept selection, and A3 architecture overviews and reports.

The amount of samples is too low to answer the main research question. The authors observed that application of methods and techniques from the body of knowledge is far from trivial. Current practice and culture in the organization constrain application of methods and techniques. In many cases, there is a clear potential for improvement. The researchers selected methods or techniques for application based on observed problems in the past. The evaluation often shows interest and a certain enthusiasm from the stakeholders for these methods and techniques. However, this enthusiasm in itself is insufficient to trigger changes; more pulling force is required to introduce these techniques at a broader scale.

Chapter 6 – Emergency lighting facilities are widely installed in public areas such as school, hospital, station, etc.

According the fire regulation in most of countries around the world, such facilities must be installed and checked regularly to ensure their normal working condition. However, usually they are suspended from a high place and distributed widely in the building.

Presently, the checking responsibility depends on human operation only, taking at least 30 mins for each device. This chapter develops a remote automatic checking system for a number of emergency lights via Internet simultaneously. The proposed system can check the status of emergency lights including the battery, charger and light using a Microprocessor-based Detector in a pre-scheduled period. Every emergency light is assigned a respective identification number.

Then, each Microprocessor-based Detector can transmit the checked results to the Nearby Computer via RS485/232. Using TCP/IP, the Remote Computer can communicate with the Nearby Computer for a long distance via Internet. Accordingly, all checked outcomes received by the Nearby Computer can be transferred to the Remote Computer for data monitoring and recording. Furthermore, the Remote Computer can send the control signals to the Microprocessor-based Detector for a remote real-time checking operation.

Experimental results confirm that the proposed system design presents a good performance in term of robust, stable, and fast response.

In: Advances in Systems Engineering Research
Editors: Elena Fermi and Adam Lamberti

ISBN: 978-1-62948-310-8
© 2013 Nova Science Publishers, Inc.

Chapter 1

PUTTING SYSTEMS ENGINEERING INTO PRACTICE: THE MODEL-BASED SYSTEMS ENGINEERING PARADIGM

***Ana Luísa Ramos**[*] **and José Vasconcelos Ferreira**[†]*
Department of Economics, Management and Industrial Engineering,
GOVCOPP Research Unit, University of Aveiro, Aveiro, Portugal

ABSTRACT

This chapter aims to explore the emerging Model-Based Systems Engineering (MBSE) paradigm as the fundamental way to put Systems Thinking and Systems Engineering (SE) into practice.

To engineer the modern large, complex, interdisciplinary systems-of-systems, the collaborative world teams must "speak" the same language and must work on the same "matter". The "matter" is the System Model and the communication mechanisms must be supported by standard, flexible, and friendly modelling languages. The evolving MBSE approach is leading the way and it is expected to become a standard practice in the Systems Engineering field in the next decade. The future of MBSE will be facilitated by the continuously evolving information technologies (computing power, storage and analysis capacities, distributed capabilities, virtual networking, etc.) as well as by the fine-tuned profile of systems engineers (the proliferation of SE courses at the various graduation levels and the adaptive profile innate to the new generations will contribute to the Systems Engineer of the future).

As a recent trend within SE field, MBSE does not have, by this time, a dedicated piece of fundamental theory. Nevertheless, there is a universal idea behind all the ongoing published research efforts: the critical importance of modelling to engineer complex socio-technical systems.

This chapter constitutes an attempt to gather the more relevant science in the topic (e.g., main definitions, standards, visual modelling languages, methodologies) and also presents a new agile methodology (LITHE - Agile Systems Modelling Engineering) to enhance the application of MBSE.

---

[*] Corresponding author e-mail: aramos@ua.pt.
[†] E-mail: josev@ua.pt.

INTRODUCTION

The contemporary world is crowded of large interdisciplinary complex systems made of other systems, personnel, hardware, software, information, processes, and facilities. An integrated holistic approach is crucial to develop these systems and take proper account of their multifaceted nature and numerous interrelationships. As the system's complexity and extent grow, the number of parties involved (i.e., stakeholders and shareholders) usually also raises, thereby bringing a considerable amount of points of view, skills, responsibilities, and interests to the interaction.

The field of Systems Engineering (SE) aims to tackle the complex and interdisciplinary whole of those socio-technical systems, thereby providing the means to enable their successful realization.

Its exploitation in our modern world is assuming an increasing relevance noticeable by emergent standards, scientific journals and papers, international conferences, and post-graduate programmes in the field. This significance is probably due to the escalating complex and "hasty" nature of our present-day systems and to the interest in achieving their overall "maximum" performance through cooperative, integrative, adaptable and interoperable environments.

The challenge is getting higher as the classical systems are evolving to modern complex Systems-of-Systems (SoS) (Jamshidi, 2008; Lane and Boehm, 2008), including both technological and social perspectives (Haskins, 2008), involving a considerable component of customized services with complex human-centred aspects (Tien, 2008), and incorporating an extensive set of " ilities" like flexibility, sustainability, real time capability, adaptability, expandability, reliability, usability, and delivery of value to society (Rhodes, 2008).

Modelling is a universal technique to understand and simplify the reality through abstraction. From brain representations to computer simulations, it is difficult to find any complex (or simple) system which does not include models or which development was not based in any kind of modelling. Furthermore, "modelling is the common basis to human activities and thus its development is also a measure of our ability to understand nature, society and related issues" (Karcanias, 2004).

Model-Based Systems Engineering (MBSE) is an emerging approach in the SE field (Rhodes, 2008; Grady, 2009) and can be described as the formalized application of modelling principles, methods, languages and tools to the entire life cycle of large, complex, interdisciplinary, socio-technical systems.

This model-centric approach, which main artefact is a coherent model of the system being developed, contrasts with the traditional document-based one (Friedenthal et al., 2008). Pointed out, by Bahill and Botta (2008), as a fundamental principle of good system design, the essence of MBSE relies on the application of appropriate formal models to a given domain.

The major potential advantages of this emergent paradigm include, for instance, enhanced communications, shared understanding and knowledge capture, improved design precision and integrity, better development traceability, and reduced development risks.

In the next decade, it is expected that MBSE will play an increasing role in the practice of Systems Engineering and will extend its application domains beyond hardware and software systems, including social, economical, environmental, and human performance dimensions.

SYSTEMS ENGINEERING ESSENCE

The words which better describe a system are perhaps 'elements' (or 'parts'), 'interactions' and 'whole', and they are well stated in the mature definition of Hitchins (2003): "A system is an open set of complementary, interacting parts, with properties, capabilities and behaviours of the set emerging both from the parts and from their interactions to synthesize a unified whole". The definition by Meadows (2008) adds the 'purpose' or function, the crucial part to determine the system's performance: "a system is an interconnected set … organized in a way that achieves something".

Bahill et al. (2002) quote Rechtin's definition which is, perhaps, the most complete one: "A system is a construct or collection of different elements that together produce results not obtainable by the elements alone. The elements, or parts, can include people, hardware, software, facilities, policies, and documents; that is, all things required to produce system-level results. The results include system level qualities, properties, characteristics, functions, behaviour and performance. The value added by the system as a whole, beyond that contributed independently by the parts, is primarily created by the relationship among the parts; that is, how they are interconnected".

The issue of interaction is of critical importance for the 'systems approach' (or 'systems thinking', or 'systems view'). The challenge is to break down complexity without disregarding the mutual interdependences which impel unique emergent properties.

The importance of the 'systems view' can be formally supported by three principles of Systems Philosophy (Hitchins, 2007): i) organismic analogy (suggests that the complex systems behave like unified wholes, like the biological organisms: they have an analogous life cycle with conception, birth, growth, maturity and death), ii) holism (considers that the systems have to be viewed as wholes and the wholes are more than the sum of its parts), and iii) synthesis (assumes a combination of the complementary parts, which cannot be considered in isolation, to bring the system into existence). The emergence is the other basic tenet which argues that the properties, capabilities and behaviours of the whole cannot be explained by any of its separable parts per se but by the interactions of these parts.

This philosophy, opposite to the classical analytical Cartesian reductionism (everything can be reduced to simple indivisible parts), has guide the way to the Systems Age, to the General Systems Theory (Boulding, 1956; von Bertalanffy, 1976) and to the Systems Science, a relatively new branch of science with the first meeting of the ancestor organization of the International Society for the Systems Sciences (ISSS) taking place 60 years ago (Bailey, 2005). This broad science devoted to the study of the systems is a science of the life, of the open systems, of the wholes, of the emergence. It is "a science whose central purpose is to provide the avenue to resolution of problematic situations of whatever nature that arise from whatever source" and is a "a science that is open to imports from other disciplines and incorporates means of identifying and integrating essential components of those disciplines" (Warfield, 2003).

With multidisciplinary inputs from biology, management, psychology (in fact, the background of the leaders: Ludwig von Bertalanffy, biologist; Kenneth Boulding, economist; Anatol Rapoport, mathematician/biologist; Ralph Gerard, neurophysiologist; James Miller, psychologist; Margaret Mead, anthropologist), the systemic outputs or applications are also spread through several fields, from systems dynamics, to biology, to cybernetics, to systems engineering, to operations research, to psychology, to general philosophy (Davidz and Nightingale, 2008). The complexity and multidisciplinarity of the modern world, as well as the strong interdependencies between its elements, demand a systemic thinking. The interrelations which characterize the present networked systems cannot be ignored. Ecological networks, biological networks, social networks, electrical networks, telecommunications networks, computer networks do not work without nets, the interactions.

The Systems Engineering field can be either classified as an application of Systems Science and, consequently, its perspective is the one of the 'systems thinking' previously discussed "One could imagine a science of relationships underlying systems engineering" (Sheard and Mostashari, 2009), and as a branch of Engineering with relatively young tradition and characterized by the professional creative application of scientific principles to the design and development of systems. According to Wymore (1993), engineering is "the creative exploitation of energy, materials and information in organized systems of men, machine and environment, systems which are useful in terms of contemporary human values".

The definitions of Systems Engineering, which began to be formalized in the 1970s with the first U.S. military standard, are numerous and diverse however, they all share the underlying concepts of the 'systems approach' like holism, synthesis, interrelationships, as well as the engineering project-based ideas of system life cycle and requirements. The classical definitions, from the 1970s, are still used but their focus was mainly on the translation of requirements to design. The following ones, from the 1990s and the 2000s, are more expanded embracing a more holistic perspective, the emergent properties, and the socio-technical aspect. The definition from the International Council on Systems Engineering (INCOSE) is a leading reference since it can be understood as a consensus of the different perspectives of active members in the field "An interdisciplinary approach and means to enable the realization of successful systems. It focuses on defining customers needs and required functionality early in the development cycle, documenting requirements, and then proceeding with design synthesis and system validation while considering the complete problem".

ENGINEERING FOR COMPLEX SYSTEMS

Surprisingly, in a recent field as Systems Engineering, there are already references to "the old Systems Engineering" (or the traditional, the classical, the ordered) and "the new Systems Engineering" (Rhodes, 2008; Sheard and Mostashari, 2009). This evolution has been reflecting predominantly the nature of the systems to engineer, which in turn reflects the tremendous and continuous advances in the technological and societal fields.

This emerging meta-field of study, in a synergistically co-evolution with SE and aiming to add a broader context to the area, is called Engineering Systems: "a field of study taking an integrative holistic view of large scale, complex, technologically-enabled systems with significant enterprise level interactions and socio-technical interfaces" (Rhodes, 2008). There are some other references which label this new field as Complex Systems Engineering (Sheard and Mostashari, 2009), Engineering of Complexity (Honour, 2008), or Systems-of-Systems Engineering (Jamshidi, 2008; Lane and Boehm, 2008). The trend is to evolve to a unified systems engineering of the future.

The classical systems (the system-as-machine paradigm) were small to large scale, multidisciplinary, relatively stable and predictable, without people as component, and were typically from the aerospace and defence industries. The new ones (the system-as-organism paradigm), which must cope with the global challenges of sustainable development, are large scale, complex, adaptive, interoperable, scalable, technology intensive, human integrative and comprise, for example, the so-called "Super Systems" like transportation and sustainable energy (Hybertson and Sheard, 2008).

The perspectives of the different stakeholders, which may be conflicting and competing, must be synthesized and resolved to serve the highest order system-of-interest needs (Rhodes, 2008). So, considering the 21st century systems, the challenge to undertake is the large scale, socio-technical, complex, system-of-systems:

- *Large scale*: systems characterized by a large number of medium or small scale constituents, interrelationships, variables, uncertainties and nonlinearities; decentralized in nature and broad in scope.

- *Socio-technical*: systems with "technical works involving significant social participation, interest, and concern" (Maier and Recthin, 2002); "…include people as inherent parties of the system, are governed by organizational policies and rules and may be affected by external constraints such as national laws and regulatory policies" (Sommerville, 2007).

- *Complex*: systems "i) with many autonomous heterogeneous components (the basic building blocks are the individual agents of the system), with a boundary often hard to pin down and with no central authority; ii) displaying emergent macro level behaviour that emerges from the actions and interactions of the individual agents; the behaviour may be unpredictable with nonlinear dynamics; the agents are often organized into groups or hierarchies, in which case the structure influences the evolution of the system; such structures tend to highlight a number of different scales, any of which can affect the behaviour of the complex system; iii) self-organizing (show a decrease in entropy due to utilizing energy from the environment and to evolve from disorder and dysfunction to order and function); iv) which adapt to their environment as they evolve (they continually increase their own complexity, given a steady influx of energy (raw resources) and feedback among elements; over time, they display increasing specialization and capability; their elements change in response to imposed "pressures" from neighbouring elements" (Sheard and Mostashari, 2009).

- *Systems-of-Systems* (SoS): "man made, created and utilized to provide services in defined environments for the benefit of users and other stakeholders. These systems may be configured with one or more of the following: hardware, software, humans, processes, procedures, facilities, and naturally occurring entities. In practice, they are thought of as goods or services. The perception and definition of a particular system, its architecture and its system elements depend on an observer's interests and responsibilities. One person's system of interest can be viewed as a system element in another person's system of interest. Conversely, it can be viewed as being part of the environment of operation for another person's system of interest" (INCOSE, 2007a).

The SoS are often but not always, complex systems. The SoS demand managerial and operational independence of the component systems (Lane and Boehm, 2008). Frequently, these ones are autonomous commercial off-the-shelf (COTS) systems, with individual system life cycles, that are bought for integration into the highest level system. This integration of independent component systems usually brings ambiguity and complexity to the development.

The modern systems involve a considerable component of customized services with complex human-centred aspects (Tien, 2008), and incorporate an extensive set of challenging requirements ("ilities") like flexibility, sustainability, real time capability, adaptability, expandability, reliability, usability, and delivery of value to society (Rhodes, 2008).

The Systems Engineer is the one responsible to 'put things together', being the interface between managers, customers, suppliers, and the different specialty engineers that are part of the process development. These are usually dedicated to specific aspects of the system whereas the Systems Engineer is concerned with the integration of the pieces into a unified coherent whole, the higher system, during its entire life cycle. The success of this system depends on the 'winning perspective' of the critical stakeholders.

The traditional main roles of the Systems Engineer include: system designer, system analyst, requirements owner, information manager, coordinator and technical manager (Sheard, 1996). The current complex socio-technical challenges demand more competences to connect people, to tackle modelling tasks, and to cope with variety, holism, flexibility, scalability, and risk. A good mathematical background and strong management and communication skills are typically mandatory requirements to the function (INCOSE, 2009). The technical competences, the leadership and versatility, the capacity for systems thinking and to solve problems creatively are also part of the profile. Haskins (2007) refers to Borhaug who characterizes the 'ultimate systems engineer' as a leader and curious person with strong will, long sighted eyes, fast moving legs, long reaching arms, coordinated by a systematic and strategic brain.

The modern Systems Engineers must have a strong foundation in mathematics, statistics and fundamentals of SE, followed by core courses in 'systems design/architecture', 'systems integration', 'quality, safety and suitability', 'modelling, simulation and optimization' and 'decisions, risk and uncertainty'. The specialization courses can include, for example, 'software systems engineering', 'finance, economics, and cost estimation', 'manufacturing, production, and operations', and 'organizational leadership'. Rhodes (2006) suggests that the education of the new generation of Systems Engineers must be supported by a collaborative effort of government, industry, and academia; must evolve to a more interdisciplinary synergistic endeavour; must be influenced by a culture of experimentation and management of risk and uncertainty, and must have the ability to make decisions taking into account a broader perspective of systems' higher order capabilities.

SYSTEMS ENGINEERING BENCHMARKS

A technical standard is an established norm which allows the unified utilization of criteria, terminology, methods, processes, measures, frameworks, tools, etc. The standards are unifying references necessary to institutionalize the practice of a given discipline, helping to translate the technical perspective to a more business one, helping to clarify its relevance to society, and to meet future challenges (Arnold, 2007). Furthermore, and in emerging collaborative world environments, they facilitate the interoperability between people and organizations. The standardization is somehow a measure of the maturity, widely expansion and growing acceptance of a given field and, in this sense, Systems Engineering is still a newborn area with a lack of accepted definitions and metrics (Valerdi and Davidz, 2009).

The core set of SE standards is relatively new, with less than a decade, and is currently in intense development by the Standards Technical Committee of the INCOSE, the Subcommittee Seven of the International Organization for Standardization (ISO), the International Electromechanical Commission (IEC), the Institute of Electrical and Electronics Engineers (IEEE), and the Object Management Group (OMG). The first standards in the SE field have risen from the American military and aerospace industries, in the 1970s and 1980s, and were dedicated to the engineering process or, in other words, to "WHAT" activities are to be performed. Since then, there has been an effort to take these standards to be domain independent in order to be applicable across different sectors, and to be international.

According to Friedenthal and colleagues (2008), the taxonomy of the SE benchmarks includes five major areas: the Process, the Architecture Frameworks, the Methodologies, the Modeling Tools, and the Data/Model Interchange mechanisms.

The Process standards still constitute the predominant core of norms, being the ISO/IEC 15288: "Systems and software engineering – System life cycle processes", from 2002 (revised in 2008), the most relevant updated international benchmark. There has been a growing effort to integrate the systems and software engineering processes, along with hardware and human engineering processes due to the increasing criticality of software within systems and to the increasing emphasis on user intensive systems and value generation.

Besides the process standards, the fundamental core that provides a foundation for a Systems Engineering approach, there are other standards in the field. The Architecture Frameworks (AF) is one of those groups, which includes the standard frameworks that have been developed to support systems' (and software) architecting. According to Cloutier and Verma (2007), a framework is a logical structure or an organizational skeleton used to classify concepts, terminology, data, artefacts, etc. There are several established AF typically oriented for a given target domain. The enterprise architecting, the systems architecting and the software architecting are the classical contexts (Tang et al., 2004; Browning, 2009). In the first group we found the well-known 'Zachman Enterprise Framework', and also the TOGAF and the FEAF frameworks (Richards et al., 2007). The systems architecting has been described through the DoDAF and the MoDAF frameworks. Finally, the software architecting has been represented by the 4+1 View Model of Architecture (Krutchen, 1995) and by the more recent Model Driven Architecture (MDA), from the OMG.

The Methodologies is another group with potential upcoming benchmarks in the field. The Harmony SE, the Object-Oriented Systems Engineering Method (OOSEM), the Rational Unified Process for Systems Engineering (RUP SE), and the Object Process Methodology (OPM) are informal methodological principles that will mature and may become established norms in the next decade.

The Data/Model Interchange mechanisms support data and model exchange among tools. The UML based modelling languages have a common foundation known as OMG Meta Object Facility (MOF™) (also an ISO standard ISO/IEC 19502: 2005), an extensible integration framework for defining, manipulating and integrating metadata and data in a platform independent manner. The XMI (XML Metadata Interchange) specification, also from OMG (and also an ISO standard ISO/IEC 19503: 2005), enables the interchange of metadata between UML-based modelling tools, like UML or SysML, and MOF-based metadata repositories in distributed heterogeneous environments, through the XML (eXtensible Markup Language). Probably, the most relevant and inclusive standard in this area will be the norm STEP/ISO 10303: AP233 (Industrial automation systems and integration: Product data representation and exchange Part 233: Systems engineering data representation). Still under development, this standard is a modular vendor neutral format for interchange of systems engineering data and to support interoperability among tools.

These (formal/informal) standards constitute the core set of norms that have been driven the development of Systems Engineering. This standardization is crucial to advance the field and to establish benchmark practices across different domains.

MODEL-BASED SYSTEMS ENGINEERING PARADIGM

Model-Based Systems Engineering (MBSE) is an emerging approach in the Systems Engineering field that can be formally stated as "the formalized application of modelling to support systems requirements, design, analysis, verification, and validation activities beginning in the conceptual design phase and continuing throughout development and later life cycle stages" or "elevating models in the engineering process to a central and governing role in the specification, design, integration, validation, and operation of a system" (INCOSE, 2007b). A simplified definition of MBSE is provided by Mellor et al. (2003) "…is simply the notion that we can construct a model of a system that we can transform into the real thing". The literature refers indistinctively Model-Based Systems Engineering and Model-Driven System Design. Since the expression MBSE is more up to date and is the most used in the professional societies, it will be used in this chapter.

The main principle underlying this approach relies on the creation of a coherent model of the system being developed. This model centric approach is expected to replace, in the next years, the traditional document centric approach that is based on documents written in text. The emergence of computers in the 1950s and 1960s has strongly contributed to this paradigm shift in a considerable range of engineering disciplines like the mechanical and the electrical ones, but in the SE field the transitioning process, while becoming prevalent, is still immature (Andary and Oliver, 2007; Friedenthal et al., 2008; Rhodes, 2008).

The traditional document-based approach is typically characterized by the development of textual specifications and design documents that reflect the requirements and the system design, and are stored in paper or electronic files which are exchanged between system architects, clients, programmers, etc. The systems engineer is concerned with the control, validity, completeness, and consistency of the documents and drawings and if the system conforms to the documents. The specifications' documents for the system are depicted in a hierarchical tree and the progress of the SE activities is measured by the state of completeness of those documents. Since the system's information (requirements, design, trade off studies, engineering analysis, etc.) is scattered across several documents the access, synchronization, and traceability of the information is difficult, as well as the management of evolving versions (Friedenthal et al., 2008). As Fisher (1998) stated, the requirements are perceived as just words, the specifications are often written after the design is complete and written words are quite ambiguous.

The emergent model-based approach aims to facilitate the SE activities through the development of a unified coherent model as the main artefact. The SE process is accomplished with increasing detailed models that are all part of the system model. The major advantages of this approach include enhanced communications between the stakeholders and team members as well as a true shared understanding of the domain, improved design precision and integrity without disconnections among the representations of data, better information traceability, enhanced reuse of artefacts, and reduced development risk. As Friedenthal et al. (2008) state, "the emphasis is placed on evolving and refining the model using model based methods and tools" so, the prominence of controlling documents is now replaced by controlling the model of the system.

It is expected that this paradigm will become a standard practice in the Systems Engineering field in the next years. The standards evolution in the field, including the SysML, the ISO 10303: AP233, the XMI, the HLA, and the MDA are impelling the proliferation of MBSE. According to the INCOSE Vision for 2020 (INCOSE, 2007b), the future of SE will be model based, embracing high fidelity static and dynamic models at different levels of abstraction. The MBSE approach will expand its boundaries and all the application domains (defence, industrial, pharmaceutical & healthcare, transportation, telecommunications, energy, etc.) will be potential targets for a model-based development.

The System Model is the main artefact of MBSE and is typically developed in a modelling language, available in a modelling tool (for example, SysML in Artisan Studio, OPDs/OPL in OPCAT), depicted on graphical diagrams, and contained in a model repository. This integrated model repository so that "everyone draws from the same well" (Browning, 2009) will embrace all the relevant information for the system and will enable marketing research, decision analysis, environmental impact analysis, social and economical modelling, biological modelling, and other appropriate analyses.

The System Model is made by interconnected modelling elements that represent the key aspects of the system namely, its requirements, its structure, its behaviour, and its parametrics (Friedenthal et al., 2008). This integrated specification is usually in interaction with other engineering models (e.g., simulation models, analysis models, hardware models) that address multiple aspects of the systems, originating a complete coherent development environment. This environment is, nowadays, a global one without physical barriers and geographical constraints. Consequently, the collaborative world teams must "speak" the same language and must work on the same "matter" that, in a MBSE approach, corresponds to the System Model.

The main potential advantages of adopting a MBSE approach are referred in the literature by several authors (Fisher, 1998; Oliver, 1998; Baker et al., 2000; Ogren, 2000; Friedenthal et al., 2008; Browning, 2009) and can be summarized as the following:

- enhanced communications between stakeholders through expressiveness and rigour: the models can express complex information in ways that are easily understood by people with different skills and competencies (systems engineers, customers, software/hardware vendors, programmers, etc.), and they provide unambiguous definitions of structure, behaviour, and capability; this clearness will increase the probability of the customers to obtain what they need;

- increased ability to manage system complexity: the system model can be viewed from multiple perspectives and can be used to evaluate the impact of changes;
- improved product quality: the model ensures the compliance between the system performance and the stakeholders' expectations and can be tested for consistency, integrity, correctness, and completeness; the traceability between requirements, design, analysis, and testing is more rigorous;
- enhanced knowledge capture and reuse of information: the capture of information based on standardize methods and the corresponding abstraction mechanisms will provide a common understanding between all the involved parties and will enable the reutilization of information/modelling elements to support design evolution;
- increased productivity: the reuse of system elements and models supports design evolution reducing cycle times; the integration and testing of the system's elements is expected to present fewer errors and take fewer time; the tools provide automated document generation;
- reduced development risk: the requirements engineering and the design V&V that are done during the system's development reduce the risk of noncompliance; the system development costs can be more accurately estimated;
- improved ability to teach and learn systems engineering fundamentals: the utilization of unambiguous and clear (and standardize) representations of SE concepts enhances the teaching/learning activity.

All these advantages are mandatory to cope with the complexity of the global development environment of modern systems. This environment demands adaptive and accurate communication mechanisms that can support considerable dimension and interdisciplinarity, geographically dispersed teams, people and technology as inherent parties of the systems, cooperation and concurrency of different subsystems (that, many times, are developed by different organizations and used by different clients), the integration of legacy and COTS systems, and "personalized" standards and system descriptions. The coexistence of these features and their integration along with "the system's big picture" can be enabled by a MBSE approach. Particular care must be taken in order to ensure that completeness, integration, and synchronization are aligned with focus and simplicity ("managers prefer simple models that they understand and trust, to more realistic ones" (Little, 1970)). The transitioning to MBSE implies a considerable investment in processes, methods, tools and, obviously, in training (Friedenthal et al., 2008). The MBSE approach requires a new way of thinking and a new set of skills. The community working with the modelling tools and languages must include language/tool experts that will develop the system model and that are able to train other team members.

The MBSE metrics are the quantified values of given attributes compared to what it is expected. These metrics can be used to assess design quality, development progress, and risk and they provide an indication if the process is moving in a successful way in order to achieve a successful outcome. The metrics to evaluate the design quality embrace, typically, the satisfaction of requirements (which requirements have been allocated, which requirements have been verified, etc.), the critical performance properties (performance or physical properties of the system to be monitored such as reliability, weight, etc.), and the partitioning of the design (the level of cohesion and coupling of the design; the cohesion can be measured in terms of how a component encapsulates behaviour without access to external data, and the coupling can be measured in terms of the number of interfaces between parts). The development progress can be assessed, for example, by the number of use case scenarios completed, by the number of requirements satisfied, by the percentage of logical components that have been allocated to physical components, by the completeness of the specification of interfaces and properties, and by the number of test cases and verification procedures that have been accomplished. The development effort and risk can be managed through the COSYSMO model that aims to accurately estimate the time and effort associated with the SE activities (Valerdi, 2008). In a MBSE environment, the size drivers that estimate the magnitude of the effort can be identified in terms of number of modelling constructs such as requirements, use cases, states, test cases, system and component interfaces (Friedenthal et al., 2008).

The future of MBSE will be facilitated by the continuously evolving information technologies (computing power, storage and analysis capacities, distributed capabilities, virtual networking, etc.) as well as by the fine-tuned profile of the systems engineers (the proliferation of SE courses at the various graduation levels and the adaptive profile innate to the new generations will contribute to the Systems Engineer of the future).

Besides the hardware and software modelling integration, MBSE will extend its domains beyond engineering and will support effects-based modelling in order to enable the evaluation of the system impacts on the broader environment (social, economical, environmental, political, human behavioural, etc.). This multiple domain approach, based on a integrated, distributed and secure model repository, is expected to be accomplished by 2025 when the MBSE paradigm will be institutionalized across academia and industry.

The proliferation of domain specific modelling languages (for example, a SysML profile for traffic & environment systems) and the extensive use of model libraries will enable the systems engineer to focus on the problem particularities and to reduce development life cycles. The integrated model repository will incite the establishment of world scale collaborative environments across academia, industry, and government. It is expected that virtual simulation/visualization environments will provide realistic verification of requirements without the need to develop physical prototypes and will enable the analysis of impacts in a realistic way.

The Human-System Integration (HSI) will be another critical research aspect in MBSE (Meilich, 2008). Since people will be fundamental elements of the system and will affect its emergent behaviour, it is mandatory to establish a bridge between cognitive engineers (and all the HSI domains) and systems engineers. The HSI MBSE Initiative (from INCOSE) is currently engaged in this challenging topic of creating a "human centric" MBSE.

The evolving Model Driven Architecture® Initiative (from OMG) aims to separate the concerns, through the utilization of different viewpoint models. The utilization of the MDA in a MBSE environment is expected to improve the efficiency of SE projects in 10 20% (Cloutier, 2008). This author suggests that in a MDA MBSE approach, the Computation Independent Model (CIM) will serve to define the WHAT (the system would do), capturing and modelling the ConOps for the system, its goals, its requirements and its interactions with other systems.

The resulting high level diagrams (e.g., requirements, use cases, sequence SysML diagrams) will then be transformed into the next level, the Platform Independent Model (PIM) that defines the HOW (the system performs). The PIM will represent the system's architecture and the allocation of customer requirements to the system requirements, detailing the model of the system (e.g., block definition diagrams, allocation tables, state machines, parametric diagrams). Then, the implementation details, with the corresponding particular type of platform such as .NET or J2EE (according to the OMG, a platform is a set of subsystems and technologies that provide a coherent set of functionality through interfaces and specified usage patterns), are specified in the Platform Specific Model (PSM). This PSM is then transformed into code such as Java or C#. The UML/SysML models are the main source of code production. The AndroMDA, the OpenMDX, and the Eclipse Modeling Framework projects are promising open source initiatives that enable the automatically conversion of UML models into deployable components for a specific platform.

The Executable UML Foundation (xUML, also developed by OMG along with the MDA) introduces action semantics for UML 2 (Mellor and Balcer, 2002). Being SysML build on UML, an executable SysML model is a future opportunity that may support simulation techniques for engineering analysis.

As obvious, these remarkable trends are potential and can only be effective if the required cultural and technical challenges will be overcome. The market forces and the field visionaries must "push the envelope to demonstrate value, exploiting opportunities and setting an example for others to follow" (INCOSE, 2007b). Some concrete recommendations of INCOSE for advancing MBSE include the development of metrics and value analysis for MBSE, the promotion of the use of modelling tools and interoperability support/standards, the identification of MBSE best practices, the advancement of standards such as SysML and AP233, the integration of SysML and simulation standards, the sharing of knowledge across the SE/MBSE community, and the development of a MBSE certification in education.

MODEL-BASED SYSTEMS ENGINEERING ENVIRONMENT

A MBSE environment includes several interacting elements (e.g., standards, formalisms, methods, modelling tools, applications, etc.) that must work together to achieve the main result, which corresponds to the development of a successful system (Ramos et al., 2012). This success is measured by the fulfilment of the stakeholders' expectations. This integrated vision is illustrated in Figure 1.

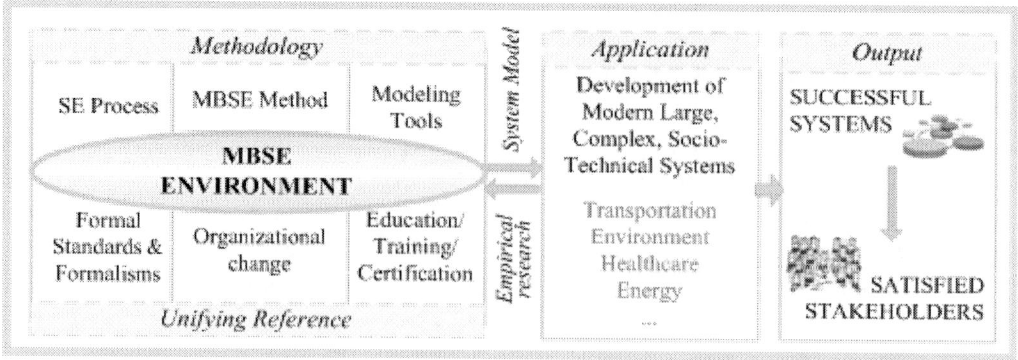

Figure 1. An integrated MBSE environment.

A methodology can be seen as "a set of related activities, techniques, and conventions that implement one or more processes and is generally supported by a set of tools" (Friedenthal et al., 2008). According to Estefan (2009), a MBSE methodology is a set of related processes, methods, and tools used to support the discipline of Systems Engineering in a model-based context.

The process is the set of interacting activities that transform the inputs into outputs (the WHAT activities to perform), the method specifies the techniques for performing the tasks of the process (the HOW to execute), and the tools are the resources applied to the method in order to improve the efficiency of the tasks (enhancing the WHAT and the HOW) (Ramos et al., 2012).

A MBSE methodology gathers all these pieces, implementing a given process, which is supported by a given method, which is enhanced by a set of tools. The capabilities and limitations of the surrounding environment, including the organizations, the technologies and the people, enable or disable the methodology and the resulting success or failure of the system's development.

The modelling tools comprise the common representations used to describe a system (concepts, attributes, structure, behaviour, entities, interactions, environment, etc.). The modelling techniques used in the field have always been predominantly qualitative and based on a graphical or pictorial representation, requiring a corresponding describing language (graphical modelling language or visual modelling language) that involves semantics (set of symbols or signs which form the basis of representations) and syntax (the proper ways of combining the symbols and signs to form thoughts and concepts). The classical representations have been, for many years, the Functional Flow Block Diagrams (FFBDs), developed in the 1950s, and with a wide spread use within the community, the Structured Analysis and Design Technique (SADT™), developed during the 1970s, and the Integration DEFinition for Function Modeling (IDEF0), a graphical notation belonging to the IDEF suite of modelling approaches and derived from the SADT, developed in 1993 by the National Institute of Standards and Technology (NIST). Other tools include, for example, the N Squared Charts, the State Transition Diagrams, and the Petri Nets.

These traditional functional decomposition representations are being "replaced" by object-oriented approaches. The modern object-oriented practices, with its roots in Software Engineering, are now pervasive in the Systems Engineering field. The characteristics of the software systems are different from those of the systems of SE (that may also include software components) and consequently, the established software oriented Unified Modeling Language (UML) lacks support from aspects like the whole/part decomposition, the requirements engineering, the performance analysis, or the trade studies. In order to incorporate these and other features, the Object Management Group (OMG) and the International Council on Systems Engineering (INCOSE) have joined efforts and developed an extension of UML for Systems Engineering: the Systems Modeling Language (SysML) (Friedenthal et al., 2009), which has been released in 2007.

This graphical language for modelling systems, which supports the specification, analysis, design, and verification of complex systems, is considered as the next *de facto* modelling language for SE.

The SysML modelling tools usually store the user model as structured data in a model repository and the model enters and retrieves that information by using the graphical representation that is, the diagrams. As Oliver et al. (2009) state "SysML continues to lack a few of the needed concepts, but has extended others in useful ways beyond historic systems engineering practice".

As UML, the modelling language for systems engineering is not attached to any methodology. The SysML also supports model and data interchange via the XML (eXtensible Markup Language) Metadata Interchange (XMI) and via the evolving neutral ISO AP233 standard.

The Object-Process Methodology (OPM), founded by Dov Dori in 2002, and the corresponding graphical and textual representations, OPDs (Object Process Diagrams) and OPL (Object Process Language), enlarge the domain of object-oriented modelling tools for SE. The provided bimodality (graphical and textual) facilitates the understanding of complexity since it is very similar to the power of both sides of the brain, the right side that acts like the visual interpreter and the left side that acts like the language interpreter. According to Grobshtein and Dori (2009), this intuitive dual notation provides a single model that is comprehensible to the different stakeholders (both technical and non-technical) involved in the development process. They are available at the software environment OPCAT (Object Process CASE Tool).

According to its author (Dori, 2002), OPM "is a comprehensive novel approach to systems engineering. Integrating function, structure and behaviour in a single, unifying model, OPM significantly extends the system modelling capabilities of current object oriented methods".

The SysML and the OPDs/OPL are the current state-of-the-art systems modelling languages. They are considerably different in terms of size and complexity. Being SysML a more "institutionalized/standardized" language with the support of the OMG and the INCOSE, and the OPDs/OPL a more intuitive simpler language with less training effort, it seems interesting to develop an hybrid approach combining the advantages of both languages and creating synergies between them (Grobshtein and Dori, 2009). This integration can strongly contribute to a common understanding of the system and to improved communications between different stakeholders, as well as to a proficient Systems Engineering collaborative development environment (Gomes et al., 2011) (some cornerstones of the MBSE paradigm).

The Harmony SE, the Object Oriented Systems Engineering Method (OOSEM), the Rational Unified Process for Systems Engineering (RUP SE), the Vitech MBSE Methodology, and the Object Process Methodology (OPM) are the main existing informal MBSE methodological principles documented by Estefan (2009).

Analyzing these methodologies one can see that they are particularly focused on the implementation of the Concept & Development phases of the Systems Engineering process. In fact, it is in these stages that SE (and MBSE) can provide considerably value added. The referred methodologies are devoted to define how to execute the functions of the process, highlighting the central role of "model the system" and considering the System Model as the main output of the design process. Table 1 summarizes the existing MBSE methodologies with its main distinctive features.

The presented MBSE methodologies are not covered by formal standards but, they will emerge as they prove their value (Friedenthal et al., 2008).

The Applications of the MBSE paradigm to real world scenarios are beginning to be published to the community. The scientific journals and the new dedicated conferences in the field confirm it.

Table 1. MBSE methodologies

Methodology	Origin	Main development approach	Main task flow	Predominant Modelling Language	Software Tools Support
Harmony-SE	IBM Telelogic	Consistent with Vee model (classical top-down approach). Service-request driven approach.	Requirements Analysis System Functional Analysis Design Synthesis	SysML	Rhapsody TAU
OOSEM	INCOSE	Consistent with Vee model (classical top-down approach) incorporating object-oriented concepts. Scenario-driven approach.	Analyze Stakehol. Needs Define Systems Requirements Define Logical Architecture Synthesize Alloc. Archit. Optimize & Evaluate Alternativ. Validate & Verify System	SysML	OMG SysML tools (integrated with other engineering tools)
RUP SE	IBM Rational	Consistent with the Spiral model (iterative and incremental development). Object-oriented concepts.	Inception Elaboration Construction Transition *Use case flow down activities*	UML/SysML	Rational Method Composer with RUP SE plug-in
Vitech MBSE Methodology	Vitech Corporation	Concurrent design. Incremental approach ("Onion model").	Requirements Analysis Behaviour Analysis Architecture/Synthesis Design V&V	SDL (based on the ERA model) EFFBDs	CORE
OPM	Professor Dov Dori	Object-oriented and Process-oriented approach. Reflective methodology.	Requirement specifying Analysis and designing Implementing Using & maintaining	OPDs/OPL	OPCAT

Probably, the first MBSE applications have arisen from the Defence and Aerospace industries that are typically characterized by systems of systems. The dimension and complexity of these systems, with a strong technological facet, had impel the evolution of engineering solutions to deal with cost overruns, schedule delays, technology constraints, and interoperability issues. The Bell Labs, in the 1940s, the U.S. Department of Defence, in the 1950s, and the NASA, in the 1960s, were possibly the first ones to recognize the importance of the systems engineering interdisciplinarity to manage and integrate large complex engineering projects.

The increasing complexity of these systems, with people, technologies, hardware, software, processes, and enterprises acting as interacting agents, demand the utilization of "intelligent and intuitive model-based systems engineering techniques" (Garcia Jr., 2009).

The "MBSE Challenge" team (collaboration between the INCOSE and the European Southern Observatory) is one of the most active initiatives in the application of MBSE principles to contemporary complex systems. The "Telescope Modeling" project and the "Space Systems" project, in current development, are examples that belong to this initiative. The major goals are to apply the SysML to solve the modelling problems, to demonstrate its adequacy to support MBSE, and to create modelling guidelines for future MBSE projects. The "Telescope Modeling" project involves the development of a next generation optical telescope that must provide a continuous mirror surface. The "Space Systems" project is working on the FireSat system whose mission is to detect, identify and monitor forest fires from orbit.

The project "Excavator Model", to evaluate interoperability issues between modelling & simulation, is being developed by the Georgia Institute of Technology. The project involves the integration of SysML models leveraged with conventional modelling & simulation tools like mechanical CAD, factory CAD, spreadsheets, math solvers, finite element analysis (FEA), discrete event solvers, and optimization tools (Peak et al., 2010).

The GEOSS - Global Earth Observation System-of-Systems, for monitoring and collecting information related to Earth's resources is another application example of MBSE. Rao et al. (2008) demonstrate the use of SysML to define the GEOSS architecture and the combined utilization of Colored Petri nets to develop the executable simulation model. Butterfield et al. (2008) propose a MBSE process to develop the architecture model and system specifications, emphasizing the SoS perspective.

Mandutianu and colleagues (2009) describe an example of a pilot application of the OOSEM methodology to design a space mission. The study reveals some encouraging potential benefits of using a MBSE approach such as, the improved communications among model designers and stakeholders, the consistent and complete representation of system models across different missions and phases, the reduction of errors and ambiguity, the reduction of design and maintenance costs, and the saved time and resources.

Simpkins and colleagues (2009) present a practical application of MBSE, using the Vitech methodology, to lead to an integrated and convergent solution for an automated parking system. The major benefits pointed out involve a better insight of the problem, a faster response to stakeholders' inquiries, a more rigorous traceability, and automated consistency checking and documentation.

Soyler and Diakanda (2010) propose the adoption of the MBSE holistic approach to capture the structure and behaviour of disaster management systems and to deal with their complexity. The SysML is used to realize the model-based paradigm.

The utilization of MBSE principles in the manufacturing domain is discussed by Tellioglu (2009). There are analyzed several case studies that aim to discover if the workers want to move from a document based approach to model based working. The results disclose the need to implement different modelling levels and strategies to engage domain workers in modelling activities.

Andersson and colleagues (2009) describe the lessons learned when introducing a MBSE approach, using UML/SysML, at Saab Aerosystems. The approach is considered to have high potential to improve engineering productivity and quality. However, there is a clear need to instigate effective modelling training programmes, with particular emphasis on model-based methods and tools.

Haan (2009) describes an application of MBSE to the health management field. A SysML model is used to demonstrate the potential competitive advantage of Prognostics and Health Management. As the author states "...MBSE methods are clearly applicable and should be highly sought by enterprises wishing to finesse a competitive advantage from PHM technologies".

The INCOSE MBSE Initiative is also working on the Urban Transportation field, along with the Florida Department of Transportation, using the cases of an Urban Traffic Signal and a Highway Maintenance System. These projects are quite immature and require further advances.

The importance of these diverse real world applications and the true essence of MBSE is quite highlighted at the following statement "...the specific tool, or language, or approach, is not the important thing; rather, systems engineers should model to understand the problem, and to communicate with others about the problem. If your modelling approach helps you accomplish that, it is a good thing" (Cloutier, 2009). The idea is corroborated by Rasmussen "the benefit of formal modelling is that we can finally stop being ambiguous and say exactly what we mean" (Delp et al., 2009).

A balance provided by Friedenthal (2009), resultant from experiences of pioneer applications of the MBSE approach, points out the following major guidelines for the successful implementation of a MBSE environment: the MBSE cultural change must be supported by an organizational change and continuous improvements principles; a well-defined MBSE methodology is decisive to support the development of the system model; adequate and customized training in languages, methods and tools is vital as well as continuous mentorship; pilot projects are required to test and validate the model based approach; well defined modelling purposes, objectives and scope are essential to properly manage the stakeholders' expectations which are "the most noticeable measure of the MBSE project success".

AGILE TOOL FOR MBSE PRACTICE: LITHE METHODOLOGY

A MBSE methodology has, as core part, the development of an integrated coherent System Model which represents the main artefact of the development environment. In a MBSE context, modelling is a fundamental activity to gain insight into how the world functions and create a shared vision among the involved stakeholders, to gain insight into complex systems, to do experimentation, to operate systems, and to negotiate, with conflicting parties, how the system will be deployed, to capture and state requirements and domain knowledge so that all stakeholders may understand and discuss them, to think about the design of a system, to organize, examine, filter, and manipulate information about large systems, to produce usable working products, to explore several solutions, and to master complex systems (Rumbaugh et al., 1999; Sussman, 2000; Buede, 2009).

The systems modelling languages play a fundamental role in these environments since they constitute a graphical way to handle information. According to the latest neuroanatomical and neurophysiologic studies, the graphical representations are processed by the right side of the brain (R mode) which is the integrative, nonverbal, intuitive, qualitative, holistic, creative, and visual thinking side capable to deal with complex visual elements ("A picture is worth a thousand words"). This utilization of the R mode, often neglected in engineering curriculum, can greatly benefit systems' architecting by looking at the whole and the synthesis. As Senge (1990) states, "If we want to see system wide, we need a language of interrelationships [R mode]". For these reasons, the authors believe that is fundamental to explore Systems Engineering with a model-based approach and with the support of graphical modelling languages such as SysML and OPM.

According to the authors, the fundamental research lines in the MBSE field will be related with (i) the development of simple and agile MBSE methodologies, and (ii) the effective utilization of graphical modelling languages able to support collaborative development environments and successful stakeholders' communication/interactions thus, successful systems (Ramos et al., 2012). As referred, a successful MBSE environment uses a given methodology to develop the model of a system and transform this model into a real thing. The methodology consists of a Systems Engineering process, a method, and a set of modelling tools that enhance the process and the method. The main artefact of a MBSE methodology is the System Model.

Regarding these aspects, and analyzing the existing MBSE methodologies, the authors believe that they are quite complex, intricate, and reveal some shortages in terms of utilization of standard modelling languages and incorporation of Human-Systems Integration concerns. In this context, and in order to overcome these limitations, it is proposed a new methodology with simple, lean and customizable processes, methods and tools to enable the widen utilization of MBSE practices.

This new methodology, the LITHE (Agile Systems Modeling Engineering), intends to be a more agile methodology using a universal and intuitive SE process, reducing the complexity and intricacy of the supporting method, emphasizing the agile principles such as continuous communication, feedback and stakeholders' involvement, short iterations and rapid response, and rousing the utilization of a coherent System Model developed through the benchmark SE graphical modelling languages (SysML and OPM). Aiming to support the development of successful systems, which satisfy the stakeholders' expectations, the methodology is particularly concerned with Human-Systems Integration.

The process (WHAT to do) is the set of activities to perform in order to transform the inputs into outputs. The SE field has a dedicated international standard (ISO/IEC 15288) that defines "the set of processes to perform during the life cycle of man made systems configured with one or more of the following: hardware, software, data, humans, processes, procedures, facilities, materials and occurring entities". From the perspective of the authors, this standard is very complete providing several enterprise, agreement, project and technical processes but it suffers from some excess of flexibility and some lack of "glue" (Ramos et al., 2010). The SE process must be intuitive, logical, universal and easy to use and to tailor. The SIMILAR process model is a simpler, intuitive, integrated and guided model, more closely related to human thinking (Bahill and Gissing, 1998), gathering the consensus of a representative group of senior systems engineers and the preference of the authors. The acronym stands for State the problem, Investigate alternatives, Model the system, Integrate, Launch the system, Assess performance, and Re-evaluate. These functions are performed in a parallel and iterative manner.

The proposed LITHE methodology encloses a revised and contextualized (in the framework of the international SE process standard) version of this SIMILAR process model. This revised version, that can be found in Ramos et al. (2010), includes a mapping of the ISO/IEC 15288 processes to the SIMILAR process in order to integrate the two models in one consistent description of the SE process. It also provides a comprehensive description of the key functions of the process. This revised version intends to be a more integrated, oriented, and "glued" version of the international standard able to be tailored by the user and easily applied in the development of modern systems.

Since every system is unique and there is no universal "recipe" to engineer them, the renewed SIMILAR process is domain independent and flexible enough to be tailored to the distinctiveness of each system while keeping the universality provided by reference international standards.

The MBSE method (HOW to do) specifies how to execute the functions of the SE process and relies on the development of a coherent model of the system. The method allied to the LITHE methodology is an agile method able to cope with the existing dynamic environments thus, based on an iterative and incremental development approach and on the state-of-the-art modelling tools (SysML and OPM). The proposed method is a combined approach of the three methods (OOSEM, RUP SE and OPM) that better satisfy the requirements. The Harmony SE method is, in a certain way, covered by OOSEM and RUP SE and the Vitech method is supported by a proposed SDL that is not coherent with the required standard modelling languages so, these methods were discarded. The LITHE methodology (Ramos et al., 2013) is depicted in Figure 2.

For each of the functions State the problem, Investigate alternatives, Integrate and Launch the system the methodology offers a simple method that provides a systematic approach to support the different stages of the process and includes, explicitly, the Human-Systems Integration concerns. Given that SE deals with socio technical systems, this aspect is crucial to attain a usable successful system.

The activities within each function should be performed iteratively, with successive refinements, like in a spiral development approach, with continuous and concurrent support provided by the transversal functions Assess performance (this function evaluates the performance through continuous monitoring, data collection, and defined metrics, and then maintains, changes, corrects, upgrades, or improves the system) and Re-evaluate (this function intends to observe the outputs in order to use this information to modify the system, its inputs and the processes or, in other word, this function provides feedback).

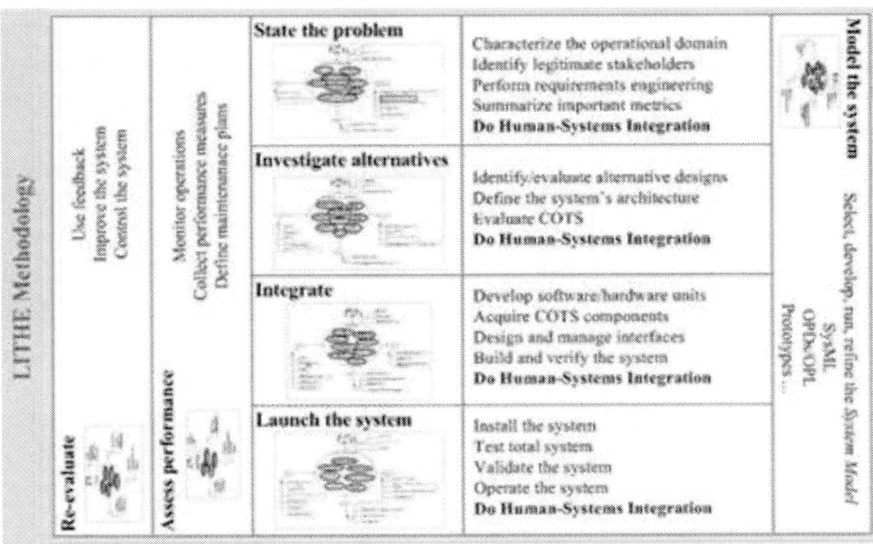

Figure 2. LITHE methodology.

The Model the system function is also transversal and is a key function of the MBSE methodology providing the modelling support that enables the development of the System Model. For each function of the process the method is applied iteratively through models (all part of the System Model) that act as working platforms among the involved stakeholders.

The State the problem function intends to identify the "reason to do" (or the problem/opportunity to address) followed by the high level description of the main functionalities of the system with all of the requirements that must be satisfied and the general performance measures agreed by the various stakeholders. It is supported by the following iterative method: characterize the operational domain, identify legitimate stakeholders, perform requirements engineering, summarize important metrics, and do Human Systems Integration. This function is supported by the three transversal functions.

The Investigate alternatives function aims to explore alternative concepts for the system's final solution and to design its baseline architecture that is, the system's elements, their characteristics, their arrangement, and their interactions. The architecture establishes a framework for the development of the system that will satisfy the requirements and its definition is one of the systems engineer most critical and creative tasks. This function is supported by the following incremental and iterative method: identify/evaluate alternative designs, define the system's architecture, evaluate COTS (commercial off-the-shelf) components, and do Human Systems Integration. It is also supported by the three transversal functions.

The Integrate function intends to bring things together so they work as a whole (system) and produce the desired emergent behaviour. This integration is done in accordance with the architectural design and the integration strategy. This function is supported by the following method: develop software/hardware units, acquire COTS components, design and manage interfaces, build and verify the system, and do Human Systems Integration. It is also supported by the three transversal functions (Model the system, Assess performance, and Re-evaluate).

The Launch the system function aims to install the developed system in its operational environment and to guarantee that the system is producing the desired outputs being working according to stakeholders' requirements. The function is supported by the following method: install the system, test total system, validate the system, operate the system, and do Human Systems Integration. Once more, the concurrent transversal functions support this function.

The "do Human Systems Integration" task is applied throughout the development process (at the main functions) by several different ways. Some examples of this critical concern entail: i) the consideration of the human dimension as an intrinsic element of the system, when stating the problem, ii) the inclusion of the organizational and social contexts in the analysis process, the placement of the key stakeholders and their goals on the centre of the process, the promotion of multidisciplinary discussion and development teams, when performing requirements engineering, iii) the consideration of usability, user-centred design, and human computer interaction principles, when developing software or hardware or bioware units (Wu and Fu, 2012), iv) the inclusion of training sessions, occupational health systems, safety, survivability and habitability concerns, when integrating systems, and v) the definition of new jobs, the establishment of new working processes and training schemes, when launching the system.

The methodology can be used horizontally (by the different alternatives of systems, elements, subsystems, and components) and vertically (by the different levels – system, elements, subsystems, components – of a well-defined system alternative). Evidently, this methodological framework is enabled/disabled and controlled by the project environment (e.g., directives and procedures, facilities, tools), by the enterprise environment (e.g., policies and procedures, standards, agreements, skills, local culture), and by the external environment (e.g., laws and regulations, available technologies, social responsibilities).

One of the major challenges in the field will be the effective utilization of graphical modelling languages able to support collaborative development environments and successful stakeholders' communication/interactions thus, successful systems. The proposed methodology has, as core part, the development of an integrated coherent System Model which represents the main artefact of the development environment. This Model must be developed with the support of the benchmark systems modelling languages.

These languages need to be competent to represent the different aspects of a system (structure, behaviour, requirements, parametrics, etc.) and capable to be understood be the different stakeholders, usually with different skills, points of view, responsibilities, and interests. The LITHE methodology has elected the current state-of-the-art modelling languages for Systems Engineering which are SysML and OPDs/OPL (from OPM).

These two languages are considerably different in terms of size and complexity but the synergies between them can strongly contribute to a common understanding of the system and to improved communications between different stakeholders. SysML is fairly large, rich and comprehensive, appropriated to provide a detailed description of the system, and uses a standard notation supported by several commercial tools but is cumbersome and requires significant learning efforts (usually, the non-technical stakeholders are not able to work with this language). The OPDs/OPL is a language more compact, intuitive, simple, and easy to learn and use, and is more adequate to model the high level concepts. Some automation mechanisms to convert one language into another are already being worked by Grobshtein and Dori (2009).

The prototypes are another category of models that can be interesting for the development of the System Model. These replicas of system's parts can be a very valuable tool to interact with stakeholders in order to elicit and clarify requirements, design usable interfaces, illustrate dynamic behaviour, and develop final solutions more aligned with the real needs and expectations. This need of prototypes will tend to decrease as sophisticated simulation/visualization (virtual reality) environments will have a tendency to increase.

The integration of these different kinds of models can provide significant added value to the engineering process.

CONCLUSION

The SE holistic approach is of increasingly relevance as the modern systems evolve to more intricate, interdisciplinary, socio-technical patterns. The field claims for (i) unified principles, models and terminology to support the application of SE to different domains, (ii) the application of SE to large scale global problems like sustainable development and global warming, (iii) lean/agile process sets and life cycle concepts, (iv) multiple views in the SE process like the socio technical and the political ones, and (v) convergent MBSE standards, modelling skills and domain specific modelling languages (INCOSE, 2007b).

The emergent MBSE paradigm requires a new way of thinking and a considerable investment in processes, methods, tools, and skills. These cultural and technical challenges demand a "proof of value" that can be accomplished through the widespread utilization of systems modelling languages, the availability of languages/tools experts able to develop coherent and integrated Systems Models and able to train other team members, the extension of application domains (beyond the traditional Defence and Aerospace industries) and implementation of pilot projects, the development and promotion of processes, methods, tools and interoperability standards, the identification of best practices, and the sharing of knowledge and experience across the SE/MBSE community.

We believe that the major developments (which will contribute to the establishment of a reliable MBSE unifying reference, made of formal standards, organizational culture, and high quality education/training) will be accomplished through accredited SE/MBSE centric programs and noteworthy empirical research. The centric programs, at the basic, master and doctoral levels, will be fundamental to provide systems engineers with the technical, communicational, modelling, and leading skills and competences that are critical to connect people and information, to cope with holism, flexibility, multidisciplinarity, human behaviour, scalability, and risk, and to solve problems creatively delivering value to society. This holistic education should be complemented by domain specific disciplines such as energy & environment or healthcare. The empirical research will be essential to drive the evolution of MBSE knowledge and to help to establish a coherent unifying reference. The experimental observations are fundamental to understand the real modern complex systems and they can be used to test MBSE hypotheses, to develop MBSE standards, and to create MBSE theories. It is our opinion that this empirical work will have as target domain the complex super systems that aim to deliver world sustainability. The Traffic & Environment, the Energy, and the Healthcare are examples of these large, complex and heterogeneous systems.

We are convinced that MBSE will be, in the next decade, a fundamental paradigm for the development of modern 21st century complex systems and will be crucial to support effective collaborative development environments. The main challenge will be to ensure that the System Model reflects the stakeholders' ideas and positions acting as a shared working platform, and the resulting System satisfies their expectations.

As the "engineering of implementation of common sense" (Holt, 2007), Systems Engineering advocates common sense principles such as "looking to the big picture", "considering the interactions", "clarifying comprehensively the problem at hand" and "listening and involving the stakeholders" but, put them into practice is not as easy as it can be expected and requires a true engineering process. The key is to assess the big picture and think systemically. The authors believes that a model-based development approach is an effective way to implement those principles and achieve the so desired "systems' successful realization".

REFERENCES

Andary, J. and Oliver, D. (2007). Models in Systems Engineering and Software Engineering. *INSIGHT-INCOSE Journal, 10*(3), 26-27.

Andersson, H., Herzog, E., Johansson, G. and Johansson, O. (2009). Experience from Introducing Unified Modeling Language/Systems Modeling Language at Saab Aerosystems, *Systems Engineering*, published online DOI 10.1002/sys.20156.

Arnold, S. (2007). Where Is Standardization Guiding Us?. *INSIGHT-INCOSE Journal, 10*(2), 41-43.

Bahill, A. and Gissing, B. (1998). Re-evaluating Systems Engineering Concepts Using Systems Thinking. *IEEE Transactions on Systems, Man and Cybernetics – Part C: Applications and Reviews, 28*(4), 516-527.

Bahill, A., Brown. P., Buede, D. and Martin, J. (2002). Systems Engineering Fundamentals. *INSIGHT-INCOSE Journal, 5*(1), 7-10.

Bahill, A. and Botta, R. (2008). Fundamental Principles of Good System Design. *Engineering Management Journal, 20*(4), 9-17.

Bailey, K. (2005). Fifty Years of Systems Science: Further Reflections. *Systems Research and Behavioral Science, 22*(5), 355-361.

Baker, L., Clemente, P., Cohen, B., Permenter, L., Purves, B. and Salmon, P. (2000). Foundational Concepts for Model Driven System Design. *White Paper*, INCOSE Model Driven System Design Interest Group, International Council on Systems Engineering.

Boulding, K. (1956). General Systems Theory-the skeleton of science. *Management Science, 2*, 197-208.

Browning, T. (2009). The Many Views of a Process: Toward a Process Architecture Framework for Product Development Processes. *Systems Engineering, 12*(1), 69-90.

Buede, D. (2009). *The Engineering Design of Systems-Models and Methods, 2*nd edition. New Jersey: John Wiley & Sons, Inc.

Butterfield, M., Pearlman, J. and Vickroy, S. (2008). A System-of-Systems Engineering GEOSS: Architectural Approach, *IEEE Systems Journal, 2*(3), 321-332.

Cloutier, R. (2008). Model Driven Architecture for Systems Engineering. *Proceedings of the Conference on Systems Engineering Research* (CSER), Los Angeles: USA.

Cloutier, R. (2009). Introduction to this Special Edition on Model-based Systems Engineering. *INSIGHT-INCOSE Journal, 12*(4), 7-8.

Cloutier, R. and Verma, D. (2007). Applying the Concept of Patterns to Systems Architecture. *Systems Engineering, 10*(2), 138-154.

Davidz, H. and Nightingale, D. (2008). Enabling Systems Thinking To Accelerate the Development of Senior Systems Engineers. *Systems Engineering, 11*(1), 1-9.

Delp, C., Cooney, L., Dutenhoffer, C., Gostelow, R., Jackson, M., Wilkerson, M., Kahn, T. and Piggott, S. (2009). The Challenge of Model-based Systems Engineering for Space Systems, Year 2. *INSIGHT-INCOSE Journal, 12*(4), 36-39.

Dori, D. (2002). *Object-Process Methodology: A Holistic Systems Paradigm*. New York: Springer.

Estefan, J. (2009). MBSE Methodology Survey. *INSIGHT-INCOSE Journal, 12*(4), 16-18.

Fisher, J. (1998). Model-Based Systems Engineering: A New Paradigm. *INSIGHT-INCOSE Journal, 1*(3), 3-4.

Friedenthal, S., Moore, A. and Steiner, R. (2008). *A Practical Guide to SysML, The Systems Modeling Language*. Burlington: Morgan Kaufmann/OMG Press, Elsevier Inc.

Friedenthal, S. (2009). SysML: Lessons from Early Applications and Future Directions. *INSIGHT-INCOSE Journal, 12*(4), 10-12.

Friedenthal, S., Moore, A. and Steiner, R. (2009). *OMG Systems Modeling Language (OMG SysML™) Tutorial*. Object Management Group (OMG) and International Council on Systems Engineering (INCOSE). Available at http://www.omgsysml.org/ [Accessed February, 2013].

Garcia Jr., J. (2009). Executable and Integrative Whole-System Modeling via the Application of OpEMCSS and Holons for Model-Based Systems Engineering. *INSIGHT-INCOSE Journal, 12*(4), 21-23.

Gomes, R., Hoyos-Rivera, G., Willrich, R., Lima, C. and Courtiat, J-P. (2011). A Loosely Coupled Integration Environment for Collaborative Applications, *IEEE Transactions on Systems, Man and Cybernetics, Part A – Systems and Humans, 41*, 905-916.

Grady, J. (2009). Universal Architecture Description Framework. *Systems Engineering, 12*(2), 91-116.

Grobshtein, Y. and Dori, D. (2009). Creating SysML Views from an OPM Model. *Proceedings of the 2nd International Conference on Model-Based Systems Engineering MBSE'09*, Herzelya and Haifa - Israel, 36-45.

Haan, B. (2009). A Model of the Competitive Advantage of Prognostics and Health Management, *Proceedings of the Annual Reliability and Maintainability Symposium, RAMS 2009*, Texas – USA, 442-447.

Haskins, C. (2007). The State of Systems Engineering in Norway. *INSIGHT-INCOSE Journal, 10*(4), 47-48.

Haskins, C. (2008). Using Patterns to Transition Systems Engineering from a Technological to Social Context. *Systems Engineering, 11*(2), 147-155.

Hitchins, D. (2003). *Advanced Systems, Thinking, and Management*. Norwood: Artech House, Inc.

Hitchins, D. (2007). *Systems Engineering-A 21st Century Systems Methodology*. London: John Wiley & Sons, Ltd.

Holt, J. (2007). *UML for Systems Engineering: Watching the Wheels*, 2nd edition. London: IET Professional Applications of Computing Series 4.

Honour, E. (2008). Systems Engineering and Complexity. *INSIGHT-INCOSE Journal, 11*(1), 20-21.

Hybertson, D. and Sheard, S. (2008). Integrating and Unifying Old and New Systems Engineering Elements. *INSIGHT-INCOSE Journal, 11*(1), 13-16.

INCOSE (2007a). *Systems Engineering Handbook – a guide for system life cycle processes and activities*. (C. Haskins, K. Forsberg, & M. Krueger, Eds.). [published online *INCOSE-TP-2003-002-version 3.1*]: International Council on Systems Engineering.

INCOSE (2007b), *Systems Engineering Vision 2020*. [published online *INCOSE-TP-2004-004-02-version 2.03*]: Systems Engineering Vision Working Group of the International Council on Systems Engineering.

INCOSE (2009). *INCOSE Careers in Systems Engineering*. Available at http://www.incose.org/educationcareers/careersinsystemseng.aspx. [Accessed August 2013].

Jamshidi, M. (2008). System of Systems Engineering – New Challenges for the 21st Century. *IEEE Aerospace and Electronics Systems Magazine, 23*(5), 4-19.

Karcanias, N. (2004). Modelling and simulation in technological and emerging fields: emerging challenges. *White Paper*, Workshop: SIM-SERV Modelling and Simulation Challenges, Working Group Roadmap for Continuous and Hybrid Simulation, University of Patras, Greece. Available at http://www.sim-serv.com/wg_doc/WG7_General_ roadmap.pdf.

Kruchten, P. (1995). The 4+1 View Model of Architecture. *IEEE Software, 12*(6), 42-50.

Lane, J. and Boehm, B. (2008). System of Systems Lead System Integrators; Where Do They Spend Their Time and What Makes Them More or Less Efficient?. *Systems Engineering, 11*(1), 81-91.

Little, J (1970). Models and managers: The concept of a decision calculus. *Management Science, 16*(8), B466-B485.

Maier, M. and Recthin, E. (2002). *The art of systems architecting*, 2nd edition. Florida: CRC Press.

Mandutianu, S., Moshir, M. and Donahue, K. (2009). Conceptual Model for Space Mission Systems Design. *Proceedings of the 19th Annual International Symposium of INCOSE*, Singapore – Singapore.

Meadows, D. (2008). *Thinking in Systems: A Primer*. Vermont: Chelsea Green Publishing Company.

Meilich, A. (2008). Status of HI/MBSE Activity. *INCOSE MBSE Initiative*. International Council on Systems Engineering.

Mellor, S. and Balcer, M. (2002). *Executable UML: A Foundation for Model-Driven Architecture*. Boston: Addison-Wesley Professional.

Mellor, S. Clark, A. and Futagami, T. (2003). Model-Driven Development. *IEEE Software*, September/October, 14-18.

Ogren, I. (2000). On principles for model-based systems engineering. *Systems Engineering, 3*(1), 1-58.

Oliver, D. (1998). The Benefits of Model Based Engineering. *INSIGHT-INCOSE Journal, 1*(3), 13-14.

Oliver, D., Andary, J. and Frisch, H. (2009). Model-Based Systems Engineering, in *Handbook of Systems Engineering and Management*, 2nd edition, Eds. Sage. A. and Rouse W., 1361-1399, New Jersey: John Wiley & Sons, Inc.

Peak, R., Paredis, C., McGinnis, L., Friedenthal, S. and Burkhart, R. (2010). Modelling & Simulation Interoperability Team Status Update. *INCOSE International Workshop IW10*, Phoenix - USA.

Ramos, A., Ferreira, J. and Barceló, J. (2010). Revisiting the SIMILAR process to engineer the contemporary systems. *Journal of Systems Science and Systems Engineering, 19*(3), 321-350.

Ramos, A., Ferreira, J. and Barceló, J. (2012). Model-Based Systems Engineering: An Emerging Approach for Modern Systems. *IEEE Transactions on Systems, Man, and Cybernetics, Part C – Applications and Reviews, 42*(1), 101-111.

Ramos, A., Ferreira, J. and Barceló, J. (2013). LITHE: An Agile Methodology for Human-Centric Model-Based Systems Engineering. *IEEE Transactions on Systems, Man, and Cybernetics: Systems, 43*(3), 504-521.

Rao, M., Ramakrishnan, S. and Dagli, C. (2008). Modeling and Simulation of Net Centric System of Systems Using Systems Modeling Language and Colored Petri-nets: A Demonstration Using the Global Earth Observation System of Systems. *Systems Engineering, 11*(3), 203-220.

Rhodes, D. (2006). New Imperatives for Educating Systems Engineering Leaders. *INSIGHT-INCOSE Journal, 8*(2), 27-28.

Rhodes, D. (2008). Addressing Systems Engineering Challenges Through Collaborative Research. *SEARI - Systems Engineering Advancement Research Initiative*. Massachusetts Institute of Technology.

Richards, M., Shah, N., Hastings, D. and Rhodes, D. (2007). Architecture Frameworks in System Design: Motivation, Theory, and Implementation. *SEARI – Systems Engineering Advancement Research Initiative*. Massachusetts Institute of Technology.

Rumbaugh, J., Jacobson, I. and Booch, G., (1999). *The Unified Modeling Language Reference Manual*. Massachusetts: Addison Wesley Longman, Inc.

Senge, P. (1990). *The fifth discipline – The Art & Practice of the Learning Organization*. New York: Doubleday Business.

Sheard, S. (1996). Twelve systems engineering roles. *Proceedings of the 6th Annual International Symposium of INCOSE*, Boston – USA.

Sheard, S. and Mostashari, A. (2009). Principles of Complex Systems for Systems Engineering. *Systems Engineering, 12*(4), 295-311.

Simpkins, P., Kleinholz, A. and Maley, J. (2009). A Practical Application of MBSE – The Automated Parking System, *Proceedings of the 3rd Asia-Pacific Conference on Systems Engineering (APCOSE)*, Singapore.

Sommerville, I. (2007). *Software Engineering*, 8th edition. London: Pearson Education Limited.

Soyler, A. and Diakanda, S. (2010). A Model-Based Systems Engineering Approach to Capturing Disaster Management Systems, *4th Annual IEEE Systems Conference*, San Diego - USA, 283-287.

Sussman, J. (2000). *Introduction to Transportation Systems*. Massachusetts: Artech House, Inc.

Tang, A., Han, J. and Chen, P. (2004). A Comparative Analysis of Architecture Frameworks. *Technical Report SUTIT-TR2004.01*, Centre for Component Software and Enterprise Systems, School of Information Technology of Swinburne University of Technology.

Tellioglu, H. (2009). Practicing Modelling in Manufacturing, *Proceedings of the 2nd International Conference on Model Based Systems Engineering* MBSE'09, Herzelya and Haifa - Israel, 75-82.

Tien, J. (2008). On Integration and Adaptation In Complex Service Systems. *Journal of Systems Science and Systems Engineering, 17*(4), 385-415.

Valerdi, R. (2008). *The Constructive Systems Engineering Cost Model (COSYSMO): Quantifying the Costs of Systems Engineering Effort in Complex Systems*. VDM Verlag.

Valerdi, R. and Davidz, H. (2009). Empirical Research in Systems Engineering: Challenges and Opportunities of a New Frontier. *Systems Engineering, 12*(2), 169-181.

von Bertalanffy, L. (1976). *General System theory: Foundations, Development, Applications*. New York: George Braziller.

Warfield, J. (2003). A Proposal for Systems Science. *Systems Research and Behavioral Science*, 20, 507-520.

Wu, C-L. and Fu, L-C. (2012). Design and Realization of a Framework for Human-System Interaction in Smart Homes, *IEEE Transactions on Systems, Man, and Cybernetics, Part A – Systems and Humans*, 42, 15-31.

Wymore, A. (1993). *Model-Based Systems Engineering*. Florida: CRC Press.

In: Advances in Systems Engineering Research
Editors: Elena Fermi and Adam Lamberti

ISBN: 978-1-62948-310-8
© 2013 Nova Science Publishers, Inc.

Chapter 2

EDUCATING MILITARY ENGINEERS AND MANAGERS WITH A SYSTEMS PERSPECTIVE AT THE AIR FORCE INSTITUTE OF TECHNOLOGY: A BLENDED RESEARCH AND EDUCATION APPROACH

Michael W. Haas and Adedeji B. Badiru
Department of Systems Engineering and Management,
Air Force Institute of Technology, OH, US

ABSTRACT

The Systems Engineering and Management Department of the Air Force Institute of Technology (AFIT) is advancing systems engineering through the weaving of education, collaborative research, and consultation processes in its Systems Engineering Research and Analysis Group (SERAG). The SERAG integrates faculty and graduate students from three curriculums; Graduate Engineering Management, Graduate Environmental Engineering and Science, and Graduate Systems Engineering, with each having experienced and multi-disciplinary civilian and military faculty, and a variety of laboratory facilities and test environments. The SERAG's synergistic integration is yielding innovative research results providing enhanced national defense capability for platforms and processes as well as military officers and DoD civilians who carry with them increased knowledge and a systems perspective upon their graduation from AFIT. The integrative approach utilized by the SERAG enhances national defense by conceiving and developing new concepts focused on the organization, training, and equipping of the Air Force that will aid the Air Force warfighter of today and of tomorrow. This chapter details the three educational courses of study themselves, summarizes several of the system-level products of technical collaborations and consultations with external clients, and describes results of the most recent collaborative research projects of the SERAG. In addition, this chapter argues for Air Force use of a systems view when identifying assignments for graduating students. Use of the systems view for assignments would align the newly acquired knowledge of the graduating student with the needed skills of the assignment accelerating innovation and organizational transformation.

INTRODUCTION

An educated workforce is a more proficient workforce. Proficiency, in this context, relates to personnel productivity, effectiveness, and efficiency. Any attempt to achieve organizational advancement must be based on an educational foundation for all employees at the respective levels of responsibilities and duties throughout the organization.

Organizational improvement is best achieved through appropriate education and utilization of the workforce. It is the belief of the authors that the military can improve organizational effectiveness through enhanced operationally efficient use of its technically educated workforce. As an example, the Air Force has a large number of degreed industrial engineers, but they are rarely placed in assignments that can directly utilize the technical skills of their profession. This chapter describes how graduate engineers and scientists studying in the Air Force Institute of Technology's (AFIT) Department of Systems Engineering and Management, as well as their faculty, integrate educational, research, and consulting activities. This integration ensures relevancy of the research and classroom topics of study providing a unique and high-level of quality educational experience for the student and enables the potential for significant benefits to be derived from a more direct structuring of assignments that would facilitate better utilization of all military engineers. In addition, using a systems view to harness industrial skills and tools by more direct assignments throughout the Air Force will lead to a widespread transformation to improved operating practices and procedures and thus, achieving increased organizational effectiveness.

AFIT is dedicated to defense-focused graduate education. Located at Wright-Patterson Air Force Base, Ohio, AFIT includes the Air Force's Graduate School of Engineering and Management as well as its institution for technical professional continuing education. A component of Air University, AFIT is committed to providing defense-focused graduate and professional continuing education and research to sustain the technological supremacy of America's air and space forces. AFIT's Graduate School of Engineering and Management is a degree granting organization offering ABET-accredited programs that include the disciplines of engineering, science, and management. AFIT is a component of the U.S. Air Force's Air Education and Training Command. The blending of research relevant to the Air Force mission, and educational coursework in systems engineering, is innovative. Each of AFIT's graduate students in Systems Engineering, whether pursuing a M.S. or a Ph.D. are required to complete a research thesis or dissertation. The focus of each student's research is coordinated between their advisory committee and in many cases, an external sponsor or advocate. The connection with the external sponsor or advocate is critical in that it reinforces the relevancy of the research direction throughout the duration of the project. Military and civilian students, who were selected by their organizations to attend AFIT, bring with them sets of personally experienced operational issues, which form a foundation of potential research topics. At AFIT, the Systems Engineering program, one of three in the Department of Systems Engineering and Management, is a curriculum leading to a degree (MS or PhD) in systems engineering. The other programs in the department are the Graduate Engineering Management program which includes the discipline of Cost Analysis, and the Graduate Environmental Engineering and Science program which includes the discipline of Industrial Hygiene. Each of these programs are described below as well as how student research blends into the student's coursework and remains relevant to the Air Force mission.

1. AFIT'S BACKGROUND OF EDUCATIONAL EXCELLENCE

The Air Force Institute of Technology traces its roots to the early days of powered flight when it was apparent that the progress of military aviation depended upon special education in this new science. Before 1919, aviation officers were educated at the Massachusetts Institute of Technology. Then, in 1919, the Air School of Application was established at McCook Field in Dayton, Ohio, the home of Orville and Wilbur Wright. When Congress authorized creation of the Air Corps in 1926, the School was renamed the Air Corps Engineering School and moved, along with all the operations at McCook Field, to Wright Field in 1927. Shortly after Pearl Harbor, the school suspended classes, but it reopened as the Army Air Forces Engineering School in 1944 to conduct a series of accelerated courses to meet emergency requirements.

After World War II, in 1946, the Army Air Forces Institute of Technology was established as part of the Air Materiel Command and was composed of two colleges: Engineering and Maintenance, and Logistics and Procurement. These colleges were later re-designated the College of Engineering Sciences and the College of Industrial Administration. When the Air Force became a separate service in 1947, the Institute was renamed the Air Force Institute of Technology. That same year, the Air Installation Engineering Special Staff Officer's Course began. In 1948, responsibility for managing officers attending civilian institutions was transferred to AFIT.

In 1950, command jurisdiction of the Institute shifted from Air Materiel Command to Air University with headquarters at Maxwell Air Force Base, Alabama. The Institute, however, remained at what is now known as Wright-Patterson Air Force Base. In 1951, the two AFIT colleges were combined into the Resident College. The Institute established a logistics education program at Wright-Patterson Air Force Base in 1955, and The Ohio State University conducted the first courses on a contract basis. In 1958, the Air Force Institute of Technology began a series of short courses in logistics as part of the Air Force Logistics Command Education Center. Later that year, the School of Logistics became a permanent part of the Air Force Institute of Technology.

In October 1954, the Engineering Council for Professional Development accredited the undergraduate Aeronautical and Electrical Engineering programs. The college was later divided into the School of Engineering, the School of Logistics, and the School of Business. The first undergraduate engineering degrees were granted in 1956, and the first graduate degrees in business in 1958. Since then, AFIT has awarded more than 18,000 graduate degrees and 600 doctor of philosophy degrees.

The School of Business programs were transferred to civilian universities in 1960. In 1963, the School of Logistics was re-designated the School of Systems and Logistics. The Civil Engineering Center was also re-designated as the Civil Engineering School.

A year earlier, in 1962, the Minuteman On-Site Program got underway. AFIT started the first program, leading to a master's degree in aerospace engineering, at Malmstrom Air Force Base, Great Falls, Montana.

Over the next 30 years, the Institute's organization changed little, but it continued to grow and respond to the changing needs of the Air Force. New programs were developed, and others were terminated. As an example, the Institute granted its last baccalaureate degree in 1985. In 1992, the Institute reorganized from three to four resident schools by specifically removing all graduate programs from the School of Systems and Logistics and establishing a new school, the Graduate School of Logistics and Acquisition Management. On October 1, 1999, the Graduate School of Logistics and Acquisition Management and the Graduate School of Engineering were combined to become the Graduate School of Engineering and Management.

In 1995, the Air Force Institute of Technology established another program to be offered at a distant location. The Air Mobility Program, taught at Fort Dix, New Jersey, is a yearlong program designed to provide officers assigned to Air Mobility Command the opportunity to further their education in a course of instruction specifically designed to enhance their expertise as operational airlift logistics experts. The first class of 10 students entered in the spring of 1995 and graduated the following May. The program utilizes facilities located adjacent to McGuire Air Force Base, New Jersey, home of Air Mobility Command's east coast operations center. Institute instructors travel to the Fort Dix site to teach these courses.

The Institute has long been an active participant in the larger educational community. In 1967, AFIT became a member of the Dayton Miami Valley Consortium, which later changed its name to Southwestern Ohio Council for Higher Education. The Council is an association of colleges, universities, and industrial organizations in the Dayton area, which are united to promote educational advancement. The Institute has traditionally been active in both the council and in other community and inter-institutional programs. In 1995, AFIT joined with two other local institutions, Wright State University and the University of Dayton, to form a consortium called the Dayton Area Graduate Studies Institute.

This consortium's purpose is to coordinate, integrate, and leverage the resources of the three schools to improve and expand graduate-level educational opportunities in the engineering disciplines. This consortium has since expanded by adding The Ohio State University and the University of Cincinnati as affiliate members. The Ohio Board of Regents, the educational governing board for the State of Ohio, funds the consortium to provide scholarships for graduate engineering students at the local institutions. In addition, the Board of Regents provides state funds to encourage collaborative research in support of the Air Force Research Laboratory at Wright-Patterson Air Force Base.

In April 2002, after a visit to the Air Force Institute of Technology, the Secretary of the Air Force, James Roche, directed the stand up of a Center for Systems Engineering. The Center's purpose is to develop and refine Systems Engineering processes and implementation, to establish baselined best practices with the DOD industry, and academia, and to manage the development and education of Air Force acquisition and sustainment personnel by incorporating hands-on experience. Air Force senior leadership made a decision in 2012 to close the center as a formal organization. That same year, as a result of another initiative by Secretary Roche, the Air Force Institute of Technology opened its Graduate School of Engineering and Management to senior enlisted personnel, fulfilling part of his vision of equipping airmen and the civilian workforce with science, technology and systems engineering skills.

Some of the most accomplished engineers and scientists in Air Force history are alumni of the Air Force Institute of Technology. Air Force pioneers General George Kenney, General Jimmy Doolittle, and General Bernard Schriever attended Air Force Institute of Technology programs prior to the time degrees were conferred. General Lawrence Skantze, who culminated his career as the commander of Air Force Systems Command, was one of the early degree graduates, while General Lester Lyles, former Commander, Air Force Materiel

Command, is an AFIT Civilian Institutions program graduate. Several NASA Astronauts attained graduate degrees at AFIT including Major General William Anders (Back Up Pilot, Apollo 11), Colonel Guion Bluford (the first African American in space), and two of the Mercury 7 original astronauts: Colonel Gordon Cooper and Lieutenant Colonel Virgil "Gus" Grissom.

The effects of the Air Force Institute of Technology's educational programs pervade the Air Force and Department of Defense. Graduates are assigned to a wide range of positions in a rapidly changing technological environment. They become both practicing engineers and broadly educated leaders. No matter what degree a student earns, AFIT's primary goal is to graduate mission-ready men and women who can positively impact the Air Force.

As the Air Force Institute of Technology looks toward its tenth decade of operation, faculty and staff members reflect with pride on the contributions the Institute's graduates have made on engineering, science, technology, medicine, logistics, and management. These immeasurable contributions have been vital to national security. As a result, Congressman Turner pushed through legislation authorizing defense industry employees to attend AFIT. Additionally, the Air Force Cyber Technical Center of Excellence and the Office of the Secretary of Defense Scientific Test and Analysis Techniques in Test and Evaluation Center of Excellence are housed at AFIT.

The future promises to be even more challenging than the past, and AFIT is prepared to continue providing the environment and the opportunity for Air Force personnel to develop the professional and technological skills needed to master this dynamic challenge. Today, AFIT mission is stated succinctly as:

"Advance air, space, and cyberspace power for the Nation, its partners, and our armed forces by providing relevant defense-focused technical graduate and continuing education, research, and consultation."

2. EFFECTIVE PERSONNEL UTILIZATION THROUGH STRATEGIC ASSIGNMENTS

An assignment to a task implies being involved in various aspects of the task's requirements. Communication complexity is one aspect that is often overlooked in the assignment of personnel to tasks where they do not add any value. The increased requirement for multi-person and hierarchical communication is the most subtle of wastes in any organization because it not only impedes overall organizational performance, but it also demeans the utilization of critical resources.

A structural review of communication, cooperation, and coordination of personnel can identify where resource utilization improvement can be achieved. Often times, the assignment requirements within the air force are not in sync with the prevailing constraints. The resolution of such problem requires integrative communication, cooperation, and coordination.

This is achieved through a systems-based approach. Basic questions of what, who, why, how, where, and when should always be addressed strategically when doing military engineer personnel and resource assignments.

It highlights what must be done and when with respect to the following:

- Does each participant know what the objective is?
- Does each participant know his or her role in achieving the objective?
- What obstacles may prevent a participant from playing his or her role effectively?

In recent years, the Air Force has announced several long-term efficiency initiatives, many of which center on organizational transformation programs. For these initiatives to be fully successful, the Air Force must leverage the capabilities of engineers among its ranks, and educate more engineers where the needs exist. The current process for managing the utilization of scientists and engineers in the Air Force does not adequately differentiate among technical disciplines and specialties between and within career fields. Moreover, the system lacks stability in that it changes rapidly and is rarely updated. For example, Table 1 shows selected industrial engineering functional areas that can offer significant operational improvement to support Air Force missions. It is noted that:

- Industrial engineering is one of many in "62" career fields that are not properly utilized.
- The first step in dealing with the STEM (Science, Technology, Engineering, and Mathematics) manpower shortfall is to make major changes in the way we currently utilize the technical specialties (AFSB, 2010).

Table 1. Selected IE Functional Areas

| Facilities location and layout design |
| Operations planning |
| Inventory planning and control |
| Material handling and distribution |
| Technical data and information systems |
| Transportation, packaging, and handling |
| Analysis of queues and bottlenecks |

Table 2. Sample Utilization of Aeronautical and Industrial/Operations Research Engineers

	Aeronautical Engineering	Industrial Engineering/ Operations Research
Education match	51%	53%
CC & PME assigned	63%	59%
PAFSC only	78%	78%
PAFSC + 63	74%	71%

A sample of 100 officer records of "62"-officers with disciplines by education in each of the aeronautical engineering and industrial engineering & operations research fields was taken and analyzed to determine the match of assignments against respective education. The study was conducted independently by two senior officers. Each scored each personnel a "1" if the assignment is related to his/her discipline, and a "0" if not. The results presented in Table 2 indicate that personnel are routinely given assignments that are not aligned with their disciplines.

3. CASE EXAMPLE OF INDUSTRIAL ENGINEERING IN THE U.S. AIR FORCE

Industrial engineering (IE), by virtue of its legacy of efficiency, provides a strategic option for achieving organizational transformation. IE emerged in 1900 out of industrial needs for efficient work systems and processes that better utilized workers in factories and operations. The key elements of IE practice are people, processes, and products with a focus on optimum performance and continuous improvement, often in allocating scarce resources. Enterprise transformation is a coordinated pursuit of operational improvements throughout the military by way of an integrated process of applications of various tools and approaches for improving operations while reducing resource expenditures. The concept of Lean was introduced many years ago for improving manual work in factory and construction operations. The modern lean principles that have evolved from the early IE practitioners were developed for production and manufacturing systems. The applications of lean principles for service operations are quite new. Military needs are predominately for service operations. The military enterprise substantively and directly affects the national economy either through direct employment, subcontracts, military construction, or technology transfer. It is, thus, fitting to expect that military enterprise transformation can have direct impacts on general civilian enterprise transformation programs. Kotnour (2011) presents the fundamental challenges of enterprise transformation with a view toward developing a universal framework for assessing transformation effectiveness. Some of the key elements he suggested are:

- Successful change is leadership driven
- Successful change is strategy driven
- Successful change is project managed
- Successful change involves continuous learning
- Successful change involves a systematic change process
- The above elements, within the context of Air Force enterprise transformation, are all within the scope of the application of industrial engineering.

As it is with most organizations, the major generic goals of the US Air Force are to *eliminate waste,* achieve *quality improvement* in products and services, and *optimize resource utilization.* Industrial Engineers are educated in the tools and techniques required to pursue these goals. Susan Blake stated it simply, that IEs make systems function better together with less waste, better quality, and fewer resource (Blake, 2011). Existing IE techniques, modified for military implementation, can be brought to bear to achieve the Air Force's generic goals. The US military, particularly the Air Force, has a cadre of industrial engineers in the engineering career field but they have not been formally engaged nor identified as resources for implementation of enterprise transformation.

Although the case of the US Air Force is cited for discussion purposes, the transformation strategies are equally applicable to other service branches in the military. The application of industrial engineering from a systems perspective can directly contribute to the most recent priorities established by the Chief of Staff of the Air Force. These priorities are summarized below:

1. Reinvigorate the Air Force Nuclear Enterprise

    - Focus on precision and reliability in Air Force processes
    - Fulfill mandate for accountability

2. Partner with the Joint and Coalition team to win today's fight

    - Aggressively adapt Air Force ways and means across the spectrum of Command and Control (C2) and Intelligence, Surveillance, and Reconnaissance (ISR)
    - Focus on Joint capabilities, interoperability, and mutual operational trust

3. Develop and care for Airmen and their families

    - Train personnel to be ready for 21st Century challenges
    - Reinforce War-fighting Ethos and expeditionary combat mindset
    - Accommodate demands on families

4. Modernize aging air and space inventories

    - Get all services to "reset" and build for the future
    - Build a balanced force for the future

5. Achieve Acquisition Excellence

    - Focus on Process, People, and Performance

These priorities can best be achieved through a systems view of overall operations. A systems approach facilitates a comprehensive consideration of all the facets of an organization (Badiru 2006). Sustainability of operational performance depends on several factors including need pre-analysis, operational requirements, work specifications, organizational capabilities, service infrastructure, administrative processes, and process standardization. The interplay of all these factors requires an integrated systems approach.

A review of the Air Force personnel database shows that there are nearly 1,000 civilian industrial engineers, as coded by educational attainment (degrees) in industrial engineering, and over 300 military industrial engineers. The frequent re-assignments that occur in the military assignment process assures that the IE diversity and relative composition will change frequently, thereby creating challenges for a cohesive utilization of IEs in the Air Force. In such a large workforce, it becomes very critical to develop comprehensive strategies to identify and effectively utilize industrial engineers. Presently, even the selection for advanced education often follows an ad hoc process and could be done better (Switzer, 2011). Using a systems framework to identify assignments for military engineers placing industrial engineers in roles that best utilize their education, training, and skills would be in the best interests of the Air Force.

Figure 1 graphically depicts the number of USAF officers with industrial engineering degrees as a histogram of distribution by rank for 2010 Air Force assignments.

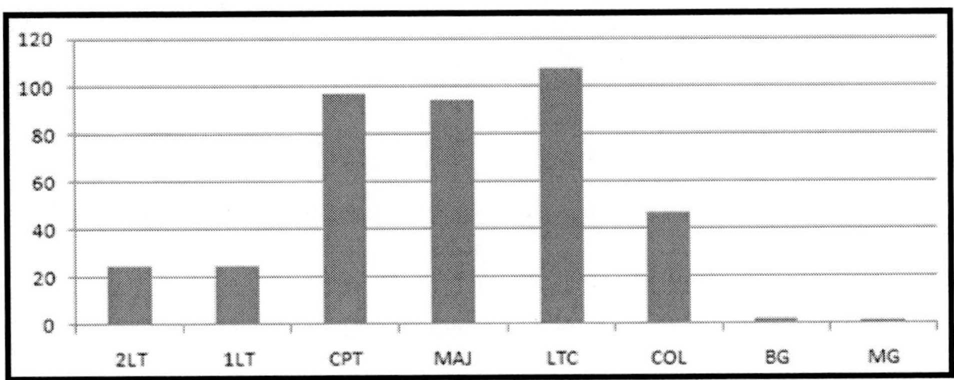

Figure 1. December 2009 Distribution of Air Force IE Assignments.

The officer assignment system tracks only the last two degrees awarded. Because of this, if an officer has a BS, MS, and PhD, only the last two degrees are in the data base. This deficiency may distort the data when the officer earns a BS in IE and then gets other degrees at the MS and PhD levels in other disciplines. However, this distortion is believed to be insignificant as it applies to only those who have earned PhDs (or professional degrees) that are non-IE. The assignment data is highly dynamic as it is affected by frequent re-assignments and the transitory status of those who have applied for retirement. Thus, the data must be tracked frequently to capture the most up-to-date situations.

At first glance, it may be surprising that there are such low numbers of military IEs in a technically-oriented workforce of over 300,000, but the high number of civilian IEs is equally surprising. In contrast to the military IEs, the civilians IEs have a specific job series identified as IE. However, many of the civilian job assignments are mis-aligned with the specific skills an IE brings to the workplace. The indifference to the specifics of the degree is revealed in assignments for IEs such as civil engineer, force support, special investigations, and others. It is believed that applying the following recommendations will aid organizational transformation achieving higher operational efficiency, better quality, and higher levels of productivity:

- Focus on total system approaches for continuous quality and efficiency improvements. Process improvement must extend beyond local commands. It must permeate throughout, and sometimes beyond, MAJCOM (Major Commands) levels to overcome suboptimal solution traps.
- Address organizational needs for maintaining and updating methods and standards, and identifying new areas for process improvements.
- Here possible, use modeling and simulation as a method to prototype and evaluate organizational process modifications before they are implemented

- Apply the analyses and recommendations above to all fields of military engineering.

Industrial Engineers, both military and civilian, are of most value to the Air Force when their graduate education is includes classroom and research exposure to topics of critical importance to the Air Force and to science and technology needs explicitly defined by Air Force senior leadership.

AFIT's Department of Systems Engineering and Management (AFIT/ENV), within the Graduate School of Engineering and Management, is uniquely organized and operating to meet these demands. The Department of Systems Engineering and Management, its programs of study, organizational structure, and recent research foci, are described in the following paragraphs.

4. SYSTEMS ENGINEERING RESEARCH AND ANALYSIS GROUP

The Systems Engineering Research & Analysis Group (SERAG) is an outreach activity providing systems-based research and analysis support for Air Force organizations.

The SERAG fully integrates the research opportunities that exist by taking a systems perspective across multiple scientific disciplines within the Department of Systems Engineering and Management with the educational mission of AFIT while providing consulting to Air Force organizations where critical needs exist. SERAG fully utilizes the expertise of the Department of Systems Engineering and Management's faculty, as well as creating collaborative opportunities across the Graduate School of Engineering and Management.

SERAG's foundational domain is Systems Engineering. Within the SERAG, systems are bounded by Chapanis' definition of systems published in 1975 "... interacting combination, at any level of complexity, of people, material, tools, machines, software, facilities, and procedures designed to work together for some common purpose."

By applying the systems engineering approach and scientific methods to problem solving, and by including disciplines such as those found in environmental engineering and science, engineering management, cost analysis, and industrial health, innovative solutions are discovered that optimize the efficiency and effectiveness of the Air Force's mission.

SERAG uses rigorous scientific methods and the analytical techniques of systems engineering to reduce cost and improve efficiency in DOD organizations as well as providing the Air Force a sustainable and innovative educational, research, and consulting capability. Figure 2 graphically depicts the SERAG and its benefits to AFIT, and to its Air Force and DoD sponsors.

Several examples of recent research efforts performed under SERAG are listed below.

- Optimizing the Intelligence Cycle to Better Defend Against Biological Weapons at the Tactical and Operational Level on the Battlefield
- Modeling the Department of Defense Acquisition Test and Evaluation Process Time
- Integration of Automated iSPARK with Optical Tracking
- The Effects of Vestibulo-Ocular Reflex (VOR) in a Vibration Environment

- Understanding the Impact of Cyber Threats and Vulnerabilities on US Space Ground Stations
- Integration of Modular Flight Algorithms into SAUS Autopilot
- Evaluation of an Innovative Technology to Treat Perchlorate—Contaminated Water

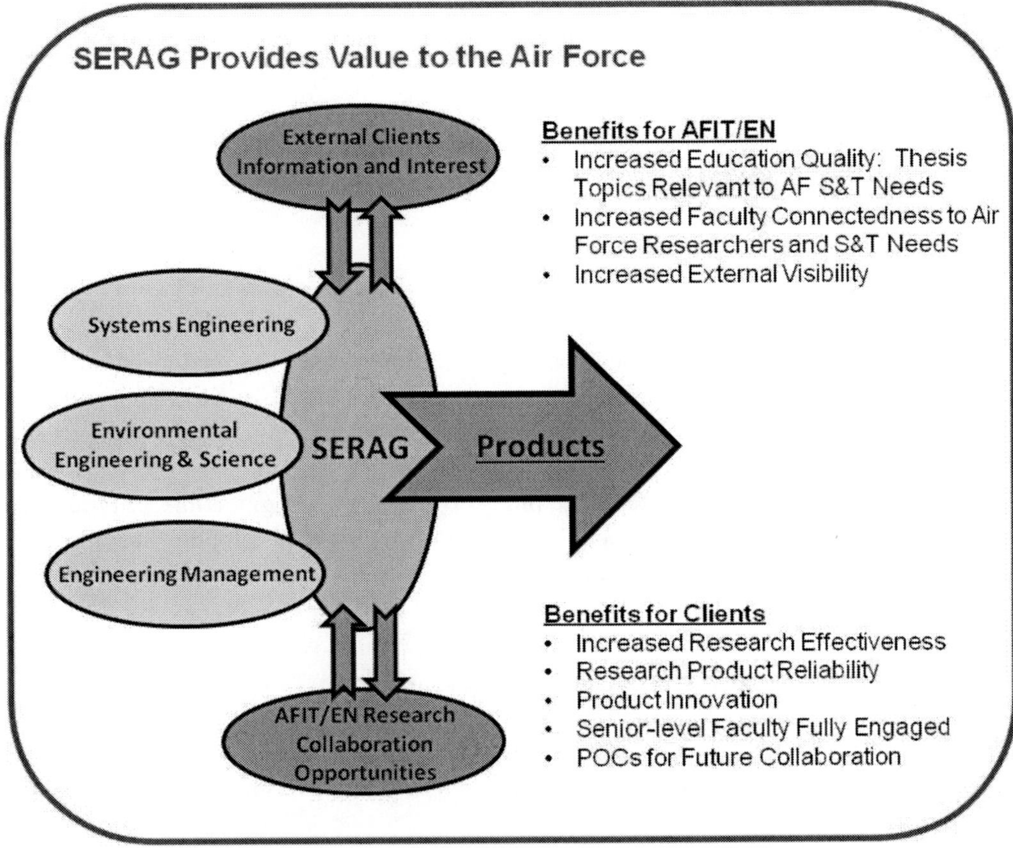

Figure 2. The Systems Engineering Research and Analysis Group.

Research performed under SERAG translates basic research understanding and concepts into prototypes and decision-quality information enabling transition into further development stages by sponsoring organizations. Several recent examples of these transitions are listed below.

- 3D Radar Displays for Air Traffic Controllers
- Systems Engineering Cost Models
- Illumination Technology for Water Purification
- Quantum Key Distribution M&S Framework
- UAS Technologies

The SERAG also transitions research concepts and results by regularly publishing in these journals:

- Journal of Nanoparticle Research
- International Journal of Critical Infrastructure Protection
- Environmental Science and Technology
- Journal of Aviation and Aerospace Perspective
- Defense Acquisition Research Journal
- Computers and Security
- Information System Security Association Journal
- Energy Policy
- Journal of Cognitive Engineering and Decision Making
- Procedia Computer Science

5. THE DEPARTMENT OF SYSTEMS ENGINEERING AND MANAGEMENT

The Department of Systems Engineering and Management is composed of faculty, graduate students and staff who participate as a team under the SERAG in approaching the research, and consultation mission of the Graduate School of Engineering and of the Air Force Institute of Technology.

In contrast to the SERAG, the disciplines of Systems Engineering, Environmental Science, and Engineering Management are separated to better support the educational mission. These three educational components are described in the following paragraphs.

5.1. Graduate Engineering Management (GEM)

Students in the graduate engineering management program are required to take a set of core courses augmented by a set of electives in one or two focus technical disciplines selected by the student.

The core courses include project management, strategic information management, engineering economics, statistics, operations research, environmental management and policy, management and behaviors in organizations, systems dynamics analysis, business process re-engineering, organizational systems, and project risk analysis.

The technical focus areas of the Graduate Engineering Management program are crisis management theory, crisis management systems, crisis project management, asset management and construction management.

The program offers electives in decision analysis, research methods, interpreting research data, and knowledge management.

5.1.1. Graduate Cost Analysis (GCA)

A program of study in Graduate Cost Analysis is paired with the Graduate Engineering management program of study. The Graduate School of Engineering and Management, Department of Systems Engineering and Management offers the Master of Science with major in Cost Analysis (GCA). The GCA program is designed to advance the knowledge and creative problem solving skills needed to effectively estimate program resources within the global military, U.S. Department of Defense (DOD) and the United States Air Force (USAF) environments. The curriculum integrates a strong foundation in quantitative concepts and techniques with specific military cost-related topics to prepare students to contribute effectively in a variety of complex and challenging roles in the global military arena. Besides the weapon system cost sequence, the curriculum includes courses in mathematical methods, quantitative decision-making, economics, cost management, risk, systems engineering, and maintenance and production management.

The minimum curriculum satisfying the degree requirements consists of two analytical methods core classes, six cost specialty courses, two system engineering courses, an economics course and twelve hours of thesis research. The analytical methods core provides a thorough background in mathematical, statistical, regression, and research methods. The cost specialty area provides depth of study in broader aspects of cost estimating. Subject areas include principles of cost estimating, advanced concepts in cost analysis, applied cost analysis, and risk analysis. The systems engineering core provides a background in system design and project management. The program culminates in an in-depth cost capstone course.

5.1.2. Focused Research Areas

One of the unique aspects of each of the systems engineering and management programs at AFIT is the emphasis on weaving the content of the student's coursework and thesis/dissertation research with issues resulting from current, and foreseen, needs of the Air Force and the Department of Defense. As examples of this connection, several recent AFIT thesis titles are provided below, many of which were sponsored externally.

Asset Management:

- Impact Of Weather Variations On Energy Consumption Efforts At U.S. Air Force Bases
- A Decision Tool To Evaluate Budgeting Methodologies For Estimating Facility Recapitalization Requirements
- Sustainable Design Policy And Leadership In Energy And Environmental Design Certification
- Selecting The Best Thermal Building Insulation Using A Multi-attribute Decision Model
- Modernizing A Preventive Maintenance Strategy For Facility And Infrastructure Maintenance
- Ground Source Heat Pumps Vs. Conventional Hvac: A Comparison Of Economic And Environmental Costs
- Construction Management:
- An Analysis Of The Design-build Delivery Approach In Air Force Military Construction
- An Analysis Of Construction Contractor Performance Evaluation System Crisis
- Management:
- A Co-citation Analysis of Crisis Management Literature
- An Exploratory Social Network Analysis Of Military And Civilian Emergency Operation Centers Focusing On Organization Structure
- Human Resource Management:

- The Impact of Operations Tempo On Intentions To Depart The Military. Does The Increase of Optempo Cause Action?
- Operations Tempo And Turnover Intentions: An Exploratory Study Of The Air Force's Explosive Ordnance Disposal Career Field
- Systems Dynamics Analysis:
- The Hybrid Counterinsurgency Strategy: System Dynamics Employed To Develop A Behavioral Model Of Joint Strategy
- Organizational Change and Development:
- Perceived Need For Change: A Test Of Individual Emotion And Contextual Influences

5.1.3. Future Research Directions: Engineering Management

In the future, the Engineering Management program will pursue the following goals:

- Understand emerging business competency requirements
  - Business Process Improvement Course added
  - Organizational Behavior class a continued anchor of the program
  - Include thesis research streams in retention and organizational change
- Define/Develop Asset Management Constructs
  - 3-course focus sequence on infrastructure developed
- Foster increased thesis-level research in infrastructure management
- Develop a more comprehensive systems perspective
  - Organizational Systems Analysis class added complementing continued Systems Dynamics class – good student feedback
  - Systems approach emphasized in crisis management focus sequence and project management classes
  - Graduate Engineering Management program co-located with the Graduate Systems Engineering program
- Maintain/bolster existing competencies required in expeditionary environments
  - Crisis Project Management class to be added
- Evaluations of case studies incorporated across curriculum
  - 85% of grade in Crisis Management focus sequence courses
  - 80% of grade in Organization Systems Analysis courses
  - ~40% of grade in Project Management Course, Systems Dynamics Course, and Business Process Improvement Course
- Incorporation of LEAN or systems perspective into education within the GEM curriculum.
- Incorporation of assignments in the curriculum that will stress and improve the critical thinking skills of the student

6. ENVIRONMENTAL ENGINEERING AND SCIENCE (GES)

The Department of Systems and Engineering Management offers the Master of Science degree with a major in Environmental Engineering and Sciences (GES). The educational objective of the GES program is that within 3 to 5 years of program completion, graduates, who would be serving as DoD officers and career professionals, will demonstrate that they can apply the principles of environmental engineering and science (encompassing air, water, soil media) to successfully accomplish their duties across the spectrum of DoD environmental consulting and management duties. These abilities may be validated through voluntary professional certification as appropriate (e.g., PE, CEM, BCEEM, QEP).

The GES degree program is accredited by the Engineering Accreditation Commission of ABET. The program was developed in coordination with, and at the specific request of, the Air Force Bioenvironmental Engineering (BEE) career field.

Program graduates are able to:

- Conduct and present methodological research to solve problems and support decisions.
- Critically analyze studies within the current literature.
- Demonstrate an in-depth knowledge within a chosen environmental emphasis area.

The GES program is conducted in six academic quarters and a short term (18 total months, beginning in September) for full time students. The short term program provides an orientation to the school and curriculum, an overview of the GES program, a review of basic mathematics and chemistry, and a refresher on written communications skills.

6.1. Graduate Industrial Hygiene (GIH)

A second program of study is paired within the GES Program. A Master of Science degree with a major in Industrial Hygiene is also offered by the Department of Systems Engineering and Management (GIH). The GIH Program was designed to provide DoD career professionals with relevant graduate education in the principles of industrial hygiene.

This GIH program was developed considering guidelines established by ABET and subject areas from the Certified Industrial Hygienist exam from the American Board of Industrial Hygiene. In addition, the GIH program was developed in coordination with the Environmental Engineering and Science (GES) degree program. Both programs were motivated by a request from the AF Bioenvironmental Engineering (BE) career field. The GES program was offered for the first time in the Fall of 2003, while the GIH program began later in the Fall of 2006. The curriculum includes department core course offerings in statistics, risk analysis, and sustainable design. The program curriculum includes course offerings in industrial hygiene anticipation/recognition/evaluation/control, radiation protection, environmental transport, epidemiology, physiology, and toxicology.

Program Educational Objectives (PEOs), which describe accomplishments that graduates should have attained within a few years after graduation, are listed below:

- Our graduates have achieved positions of leadership as occupational/environmental health consultants, instructors, or similar positions of responsibility.

- Our graduates have applied their education to address difficult technical problems for the Air Force, sister service, civilian, or foreign industrial and community environments.
- Our graduates have attained voluntary professional board certification as Certified Industrial Hygienists.

Student Outcomes (SOs), which are skills that students have after program completion, are listed below:

- The student can anticipate and recognize agents, factors, and stressors.
- The student can evaluate agents, factors, and stressors for hazard potential.
- The student can control hazards presented by agents, factors, and stressors.

6.2. GES Academics and Research

The 50-quarter hour (QH) degree requirements for the Environmental Engineering and Science degree include specified core classes plus a specialty sequence, statistics course, thesis, and electives.

The specialty sequence is intended to develop students' in-depth knowledge of a specific area of environ-mental engineering and science. A 12 QH thesis with oral defense is also required.

The GES core courses include Environmental Systems Engineering, Sustainable Life Cycle Design, Environmental Transport Processes, Environmental Risk Analysis, Environmental Sampling and Analysis, Statistics, Air Resources Management, Water Chemistry, and Water and Wastewater Treatment and Design.

Additional coursework in Public Health, Industrial Hygiene, and Epidemiology is available either through the GIH program or our affiliation with the University of Dayton and Wright State University.

These courses serve to enhance knowledge in those subjects critical to the environmental health officer. We also offer courses focusing on Chemical, Biological, Radiological, and Nuclear (CBRN) sciences including Physiologic Effects of Nuclear, Biological, and Chemical Agents and Biological Weapons Technology. Specialty sequences can also be selected from the GIH Program.

The thesis must address a real-world problem in an engineering or environmental science. Principal purposes of the thesis are to demonstrate the student's ability to integrate concepts and techniques acquired through course work and to demonstrate scholarly pursuit of a focused research question, all of which leads to enhanced capability of the graduate to pursue technical problems creatively and effectively across a broad spectrum of areas. GES students have many research opportunities. AFIT maintains close research ties with the Air Force Research Laboratory on base, as well as many Air Force and DoD organizations, civilian universities, and private organizations.

6.2.1. GIH Academics and Research

The GES and GIH programs are closely aligned and share many common core courses and available electives. The GIH program is conducted in six academic quarters and a short term (18 total months, beginning in September) for full time students. The short term program, which is the same program as what is required for GES students, provides an orientation to the school and curriculum, an overview of the GES and GIH programs, a review of basic mathematics and chemistry, and a refresher on written communications skills.

The 48-quarter hour (QH) degree requirements for the Industrial Hygiene degree will include specified department and IH program core classes. A 12 QH thesis with oral defense will also be required. The thesis can be basic or applied research, addressing a real-world problem in environment, safety, and occupational health (ESOH). In addition to degree requirements, all DoD-sponsored full-time students must complete an average of 12 credit hours per quarter over the total number of quarters in residence, for a total of 72 credit hours.

7. GRADUATE SYSTEMS ENGINEERING (GSE)

Systems Engineering (SE) is the process by which a customer's needs are satisfied through the conceptualization, design, modeling, testing, implementation, and operation of a working system. The focus on SE becomes especially important in the analysis and synthesis of large and complex systems, such as those that arise regularly in Department of Defense and the Air Force. We continually structure the program to address relevant topics, both in coursework and thesis/ research projects.

- System of Systems Engineering
- Architecture development/ analysis
- Complex adaptive systems

Over the last few decades, Systems Engineering has matured into its own discipline, with a foundation on system science using tools and repeatable processes from product development and systems engineering management. Recently, with the pervasive deployment of complex interconnected networked systems, the use of architecture has taken a central role in communicating the system of systems and enterprise-wide solutions.

6.2.2. Resident GSE Academics

The Graduate Systems Engineering (GSE) program is an ABET accredited program leading to a Master of Science (MS) degree in Systems Engineering. This program is provided to both resident students and part-time students via distance-learning (DL) modality. In residence, it is nominally a six quarter (18 month) program, with students normally entering in September and graduating in March after defending their thesis. The SE curriculum requires the following:

CORE COURSES

- Systems Engineering Design

- Engineering Software-Intensive Systems
- Systems Architecture
- Systems Engineering Management

ANALYSIS

- One of: Project Risk Analysis, Decision Analysis, Modeling and Simulation

MATH

- Intro to Probability and Statistics

ENGINEERING DEPTH

- Three or more courses in a domain, such as space systems, aeronautical systems, C4ISR, cyber systems, Test & Eval, Logistics Management, etc

DESIGN STUDY / THESIS

- Students participate in an individual or group thesis. These design projects are typically sponsored by DoD and Air Force organizations.

6.3. Distance Learning GSE Academics

The Graduate SE Program is also offered in a distance learning (DL) format. The program is nominally a 24 to 36 month program, depending on class availability and part-time course load.

6.4. Resident Intermediate Development Education (IDE) GSE Program

Air and Space Systems Engineering is one of 4 IDE programs AFIT offers for Majors selected for in-residence Professional Military Education. Curriculum follows the Graduate SE program with the exception of a Graduate Research Project (9 credit/hours) in lieu of a thesis (IDE) students arrive in May and graduate 12 months later in June. This degree program is only available to military personnel and DoD civilians selected by their service component for resident PME.

6.4. GSE Research Focus

In order to best meet the needs of our warfighting and system life cycle managing customers, discussions with joint warfighter requirements, integration, capability planning, acquisition, product and logistics centers and MAJCOM requirements organizations have focused our Systems Engineering program. For example, Systems Engineers working within the space community, should know the physics of space surveillance and/or the space environment, be knowledgeable in the wealth of unclassified and classified space technologies and systems and apply lessons learned through a sponsored space-related Capstone project. We achieve this requested focus with a set of specified electives and technical specialty courses.

Students will usually satisfy the AFIT SE degree requirements by choosing a focus area. Based on the current needs of the DoD and the Air Force, our primary focus areas are as follows:

- *Airborne Systems*: Performance and design analysis of manned and unmanned aircraft will be examined, as well as major subsystems on our aeronautical systems. These include guidance, navigation, C4, radar, propulsion and structures, and include munitions and their effects.

- *Space Systems:* Military space vehicles or satellites are not "contained" systems, because their functionally is part of a networked constellation of satellites and ground stations synergistically performing a needed mission, providing warfighter capability and creating desired effects. The extreme environments in which these systems operate necessitate unique design and development processes.

- *Cyber Warfare*: The Cyber Warfare sequence is designed to study, analyze and challenge theories on the application of cyber power (offensive and defensive) to achieve strategic and operational military objectives. Students develop technical expertise and a technical foundation to better understand and analyze communications/ networks, policy, operations, systems and technologies.

- *Logistics (supply chain) Systems*: The Logistics systems sequence is designed to provide students with graduate level education in the fundamentals of Supply Chain Management (SCM), with particular emphasis on Department of Defense (DoD) and Air Force specific applications. Statistical data analysis and basic quantitative modeling, to include linear programming, simulation analysis, and heuristics, are included. Students will be able to apply state of the art analytical and problem solving techniques to Air Force and DoD supply chain management problems.

- *Human Systems*: The Human Systems sequence is designed to provide students with graduate level education in the fundamentals of Human Factors Engineering with an understanding of Human Systems Integration, and particular emphasis on Department of Defense (DoD) applications. Courses in integration of human systems, human factors, human performance measurement, design of human computer interfaces and display/control systems, are included. This focus area will prepare students to apply state of the art analytical and problem solving techniques to design of systems having a significant human component.

CONCLUSION

The Systems Engineering and Management Department of the Air Force Institute of Technology (AFIT) is well poised to advance systems engineering through the weaving of education, collaborative research, and consultation processes. The innovative use of the SERAG as an outreach activity that encompasses the multi-disciplined faculty of the Department provides a foundation supporting the application of systems engineering principles research and analyze Air Force science and technology needs. The graduate student enrolled in programs offered by the Department of Systems Engineering and Management experience a high-quality learning environment from senior-level faculty and are exposed to research topics that are relevant to today's science and technology needs of the Air Force and the DoD. The high-quality learning environment, senior-level faculty, Air Force and DoD relevant research topics, and the outreach activities of the SERAG make the Department of Systems Engineering and Management unique within the DoD. The expertise of AFIT's graduating students can best be utilized by the Air Force for organizational transformation by explicitly aligning the student's follow-on assignment with the knowledge gain during their graduate education.

REFERENCES

AFSB (Air Force Studies Board), (2010), "Managing STEM Personnel to Meet Future STEM Needs Across the Air Force," in Examination of the US Air Force's STEM Workforce Needs, The National Academies Press, Washington, DC, 2010, pp. 75-99.

Badiru, Adedeji B. and Marlin U. Thomas, editors, *Handbook of Military Industrial Engineering*, Taylor & Francis CRC Press, 2009.

Badiru, Adedeji B., "Air Force Smart Ops for the 21st Century," *OR/MS Today*, Feb 2007, page 28.

Badiru, Adedeji B., editor; *Handbook of Industrial & Systems Engineering*, Taylor & Francis CRC Press, Boca Raton, FL, 2006.

Blake, Susan (2011), "5-minute explanation of what an IE does," *Industrial Engineering*, Vol. 43, No. 10, October 2011, p. 12.

Kotnour, Tim (2011), "An Emerging Theory of Enterprise Transformations," *Journal of Enterprise Transformation*, Vol. 1, No. 1, pp. 48.70.

Switzer, Tobias (2011), "Air Force Policy for Advanced Education: Production of Human Capital or Cheap Signals," *Air & Space Power Journal*, Winter 2011, pp. 29-42.

Thomas, Marlin U. and Adedeji Badiru, "Utilization of Industrial Engineering in the Air Force," paper presented at Air Education and Training Command (AETC) Symposium, San Antonio, TX, Jan 14-15, 2010.

In: Advances in Systems Engineering Research
Editors: Elena Fermi and Adam Lamberti

ISBN: 978-1-62948-310-8
© 2013 Nova Science Publishers, Inc.

Chapter 3

SIMULATION OF THE SEPARATION OF INDUSTRIALLY IMPORTANT HYDROCARBON MIXTURES BY DIFFERENT DISTILLATION TECHNIQUES USING MATHEMATICA©

Housam Binous[1] and Ahmed Bellagi[2]

[1]Department of Chemical Engineering, King Fahd University of Petroleum & Minerals, KSA

[2]Département de Génie Energétique (Energy Engineering Department), Ecole Nationale d'Ingénieurs de Monastir, University of Monastir, Tunisia

ABSTRACT

In this chapter the simulation of several case studies for the separation by distillation of various industrially relevant hydrocarbon mixtures are worked out. Different techniques used to separate the complex mixtures are illustrated: classical column train and enhanced distillation (extractive and reactive).

This chapter is aimed primarily at chemical engineering students. But the information presented might be also useful for postgraduate and PhD students as well as professional engineers.

An important objective of this chapter is to show how these rather complex separations can be handled with great pedagogical benefits using the computer algebra Mathematica©, a general purpose solver, not chemical engineering specific. Simulation results are compared to those obtained using the flow-sheeting software Aspen-HYSYS®.

Five case studies are worked out: the separation of natural gas liquids (NGL), the fractionation of a C_4 cut to separate 1,3-butadiene with furfural as entrainer, the production of MTBE from *i*-butene and methanol, the reverse process: production of *i*-butene and methanol by the decomposition of methyl *tert*-butyl ether (MTBE), and the equilibrium-limited metathesis of cis-2-pentene to cis-2-butene and cis-2-hexene.

For every case complete information is given such that it can be easily reproduced with the process simulator and/or alternatively worked out using the relevant Mathematica© programs available at http://demonstrations.wolfram.com.

Keywords: Enhanced distillation, NGL train, Extractive distillation, Reactive distillation, Numerical simulations, Mathematica$^{\copyright}$, Aspen HYSYS®

INTRODUCTION

Distillation is a reliable separation technique, and remains often the most efficient fractionation method. The ubiquity of the distillation installations in the gas, oil and petrochemical industry makes it indeed be considered as the "workhorse" separation process in this field.

In this chapter, the fractionation by different distillation techniques of various industrially relevant hydrocarbon mixtures is investigated. Mathematical modeling and numerical simulation are our exploration tools. The objective is to illustrate different distillation techniques used to separate complex mixtures: classical column train, and enhanced distillation (extractive and reactive).

Five case studies illustrating different separation distillation techniques are worked out.

The first case study deals with the separation of natural gas liquids (NGLs), a mixture of light hydrocarbons (ethane, propane, butane, pentane, hexane and heptanes). The raw NGLs mixture is separated using a sequence of distillation columns in which the desired hydrocarbons are gradually isolated.

The second case study is the extraction of 1,3-butadiene from a C_4-cut using extractive distillation with furfural as entrainer.

The remaining three cases are dedicated to the simulation of reactive distillation columns for:

- The production of methyl *tert*-butyl ether (MTBE) from *i*-butene and methanol,
- The production of *i*-butene by the reverse process of decomposition of MTBE,
- The equilibrium-limited metathesis of cis-2-pentene to cis-2-butene and cis-2-hexene.

The various case studies are *first* worked out using the well-known flow-sheeting software Aspen-HYSYS®. It is an important objective of this chapter however to show how these rather complex separations can also be handled with great pedagogical benefits using the more accessible computer algebra Mathematica$^{\copyright}$, a general purpose solver, not chemical engineering specific.

This approach is particularly suitable for undergraduate students, as at this level of chemical engineering formation, the synthetic pedagogical approach is more effective and can be very rewarding. The student builds all from starch and has to elaborate by himself a model for the process under investigation. During this step of model development, one learns a lot about equations of state, departure functions, activity models, MESH equations…

By integrating the basic mathematical functions with the powerful and easy-to-use programming language Mathematica$^{\copyright}$, it is possible to handle processes that would be extremely tedious to encode in traditional programming environments. Using this powerful tool let the student concentrate on the very process he/she is investigating and enhances dramatically the overall understanding of his/her project. As illustrated in this chapter various process alternatives can easily be probed (integration of the pressure drop in the stages, taking account of the non ideal efficiency of the plates, trying different solvents, etc.).

NATURAL GAS LIQUIDS, NGL, TRAIN

In a natural gas processing plant, the raw gas is pre-treated to remove liquid water and condensate, on one side, and the accompanying impurities that may interfere with the following separation processes and/or are undesirable in the final products, on the other side. Important steps in this pre-treatment are acid gas and sulfur compounds removal (CO_2, H_2S, mercaptans, etc.), dehydration and eventually mercury removal and finally nitrogen rejection if the N_2 content of the gas is high. In the next separation unit, the Demethanizer, the dry sweet gas is cooled to separate the sales gas–containing mainly methane and less than 0.1 mol% pentane and heavier hydrocarbons–from the rest, raw Natural Gas Liquids. NGL is a mixture of ethane and heavier hydrocarbons (propane, butane, pentane, hexane and heptane) and constitutes an important feedstock for petrochemical plants and facilities (steam crackers) producing olefins and aromatics. As the components of the NGL often have higher sales value compared to the gas itself [1] they are separated using a train of distillation columns in which the various hydrocarbons are boiled off one by one [2, 3, 4]. The fractionation columns are called: Deethanizer, Depropanizer, and Debutanizer (Figure 1).

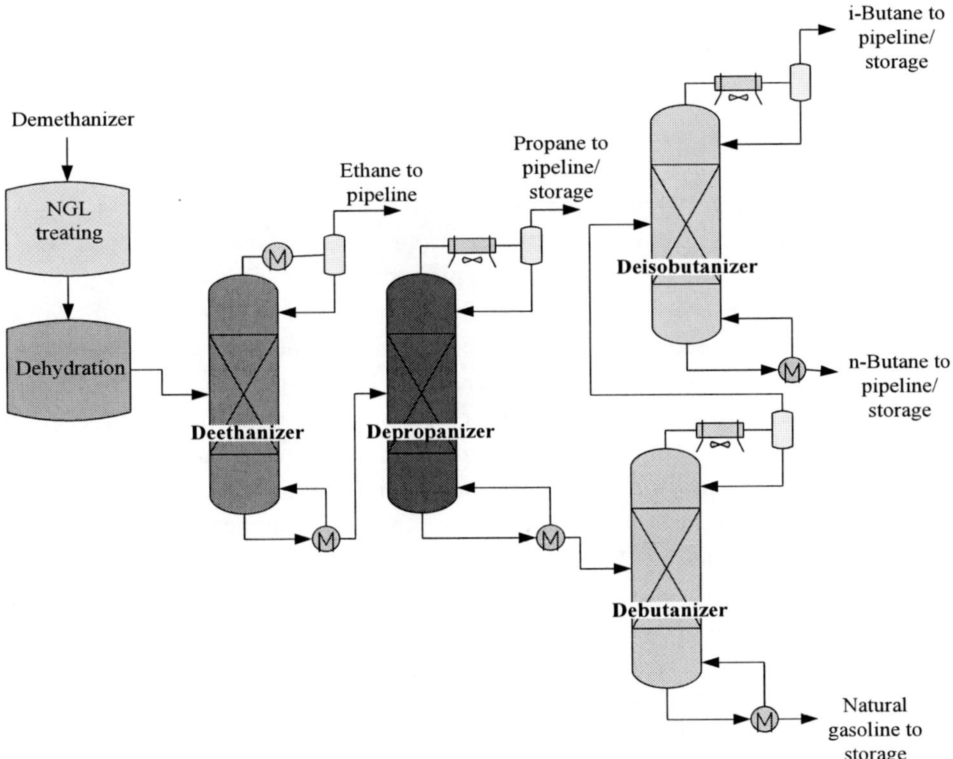

Figure 1. NGL separation train.

This first case study deals with the simulation of an NGL train. Raw NGL is to be separated in almost pure ethane, i-butane, n-butane, and a mixture composed of propane and the rest of the dissolved ethane. The bottom product of the last fractionation unit is natural gasoline.

As numerical example we consider a feed with a flow rate of 100 kmole/hr and a typical composition as given in Table 1 where C_7^+ is a pseudo-component that lumps together all components heavier than n-hexane. The four distillation columns (Deethanizer, Depropanizer, Debutanizer and Deisobutanizer) are operating at decreasing pressures: 26 bars, 17 bars, 10 bars and 7 bars, respectively. This pressure selection ensures, first, the separation of almost all components heavier than ethane in the Deethanizer at temperatures near 0°C, and allows, second, the usage of water as cooling medium in the condensers of the remaining units.

Table 1. Feed composition to the NGL train

Component	Mole Fraction
Methane	0.005
Ethane	0.370
Propane	0.260
i-butane	0.072
n-butane	0.148
n-pentane	0.085
n-hexane	0.040
C_7^*	0.020

Aspen HYSYS® Simulations

It is relatively straightforward to perform a simulation of the NGL train using Aspen HYSYS®. First, the components are chosen from the software's databank and the thermodynamic model for the prediction of the properties and the calculation of the vapor liquid equilibria—in the present case the Peng-Robinson equation of state (PR EoS, Appendix A). Then, the distillation columns for the Deethanizer, Depropanizer, Debutanizer and Deisobutanizer are added to the Process Flow Diagram (PFD, Figure 2). For the Deethanizer column, when the feed stream specifications are entered and the required column settings are selected properly, the column simulation can be performed and results are obtained. Next, in order to decrease the high pressure of the bottom stream of the Deethanizer (i.e., 26 bars), which serves as feed stream to the next column (i.e., the Depropanizer) operating at 17 bars, an expansion valve is incorporated in the PFD. Similarly, simulations can be performed for the Depropanizer and later for the rest of the distillation train (i.e., the Debutanizer and the Deisobutanizer). All distillation columns considered below have 40 equilibrium stages and in all four cases, the feed is entering at stage 20 counting from the top. The idea, behind the choice of these configurations, is to provide enough trays to separate ethane/propane in the Deethanizer, propane/i-butane in the Depropanizer, butanes/higher hydrocarbons in the Debutanizer and i-butane/n-butane in the Deisobutanizer. Furthermore, while a partial condenser is associated to the Deethanizer, a total condenser and a partial reboiler are considered in all other cases.

Results of the simulations are given in Table 2 with the composition of all material streams. We note in particular that the distillate of the Deethanizer is constituted of ethane (98.5 mol%) and methane (1.5 mol%). The top stream of the Depropanizer is composed of propane (85 mol%) with dissolved ethane (13.35 mol%) and a low fraction of *i*-butane (1.33 mol%). The steams exiting the last fractionation unit, the Deisobutanizer, are almost pure products: *i*-butane (99 mol%) in the distillate and *n*-butane (99 mol%) in the bottoms.

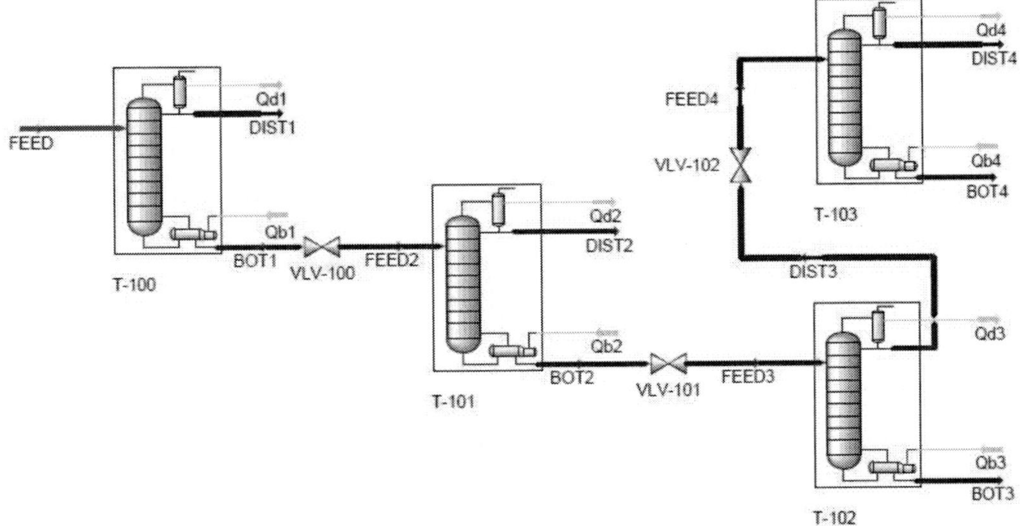

Figure 2. HYSYS® PFD of the NGL separation train.

Table 2. Results of the Aspen HYSYS® simulations

	DIST1	BOT1	DIST2	BOT2	DIST3	BOT3	DIST4	BOT4
Methane	0.0150	0.0000	0.0000	0.0000	0.0000	0.0000	0.0000	0.0000
Ethane	0.9850	0.0626	0.1365	0.0000	0.0000	0.0000	0.0000	0.0000
Propane	0.0000	0.3900	0.8500	0.0000	0.0000	0.0000	0.0001	0.0000
i-Butane	0.0000	0.1080	0.0133	0.1882	0.3149	0.0000	0.9899	0.0100
n-Butane	0.0000	0.2220	0.0002	0.4099	0.6851	0.0010	0.0100	0.9900
n-Pentane	0.0000	0.1275	0.0000	0.2355	0.0000	0.5856	0.0000	0.0000
n-Hexane	0.0000	0.0600	0.0000	0.1108	0.0000	0.2756	0.0000	0.0000
C_7^+	0.0000	0.0300	0.0000	0.0554	0.0000	0.1378	0.0000	0.0000

Mathematica© Simulations

In the four sections below, we present simulation results using Mathematica© for the same case study.

For both methods (i.e., calculations with Mathematica© and Aspen HYSYS®) and for all four separation units of the NGL train, the Peng-Robinson EoS has been used. MESH equations (Appendix B) for each one of the four separation units result in a set of 800 nonlinear algebraic equations, which are readily solved using Mathematica[©] in less than five minutes with an Intel[®] Core[™] 2 Duo CPU T9600 2.80 GHz RAM 4 GB.

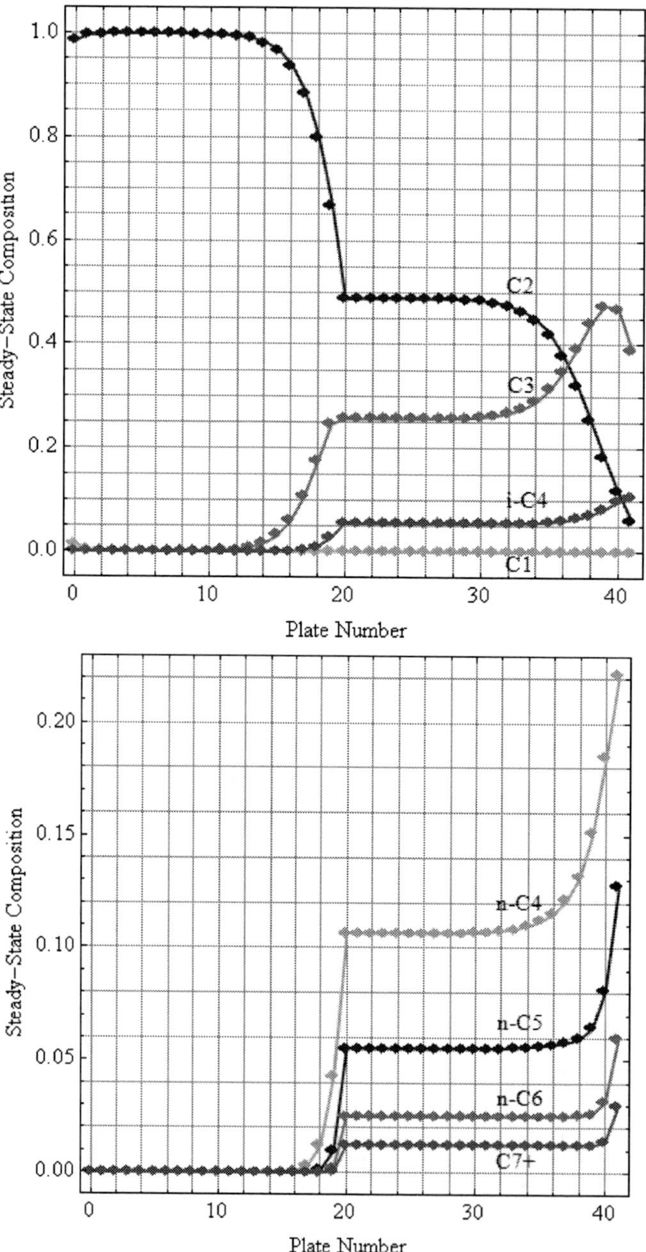

Figure 3. Steady-state composition profiles in the Deethanizer. (Continuous lines: Mathematica[©] ; markers: Aspen HYSYS[®]).

Deethanizer

In this column ethane is separated out as overhead product. The reflux ratio and the mole % of ethane in the distillate have been set equal to 3 and 98.5, respectively.

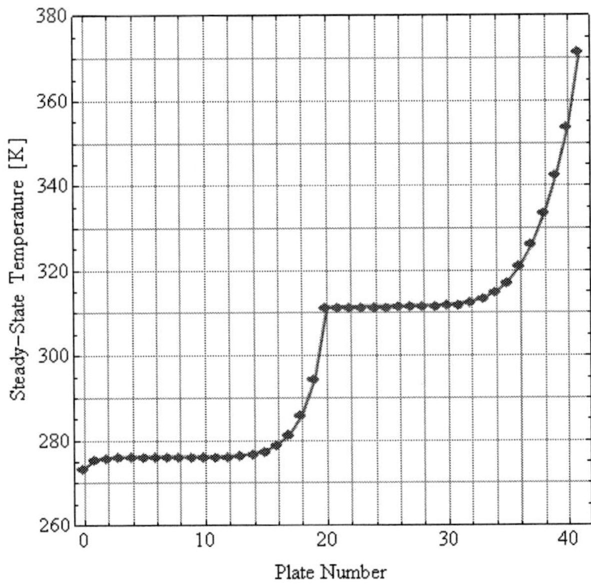

Figure 4. Steady-state temperature profile for the Deethanizer. (Continuous lines: Mathematica[©] and markers: Aspen HYSYS[®]).

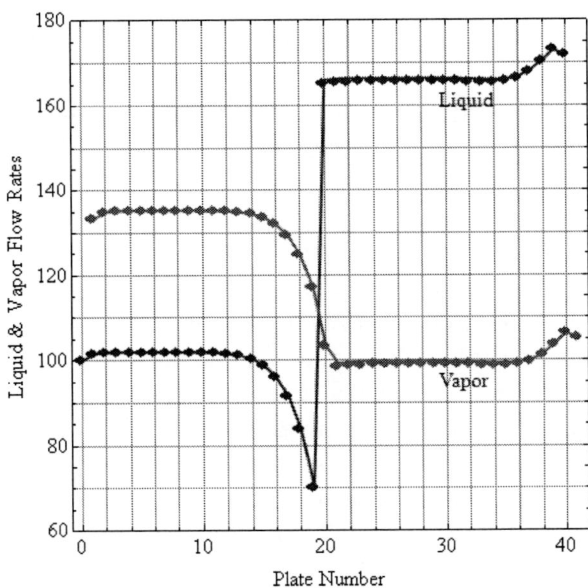

Figure 5. Steady-state liquid and vapor flow rate profiles for the Deethanizer. (Continuous lines: Mathematica[©] and markers: Aspen HYSYS[®]).

Figures 3-5 show the steady-state composition and temperature profiles as well as the liquid and vapor flow rates in the column. The distillate product has a flow rate of 33.33 kmole/hr and contains: 1.5 mol% methane and 98.5 mol% ethane. The bottom product from Deethanizer enters into the next column (i.e., the Depropanizer) with a flow rate equal to 66.66 kmole/hr and the following composition: 6.25 mol% ethane, 39.0 mol% propane, 10.8 mol% i-butane, 22.2 mol% n-butane, 12.75 mol% n-pentane, 6.0 mol% n-hexane and 3 mol% C_7^+. The cooling and heating duties of the condenser and reboiler are $1.167 \ 10^6$ and $1.542 \ 10^6$ kJ/hr, respectively.

We note from Figures 3-5 also that the results of our Mathematica$^©$ simulations (continuous lines) match perfectly the results obtained with Aspen HYSYS$^®$ (markers). This is true for the composition profiles of all of the constituents in the column (Figure 3) as well as for the temperature evolution (Figure 4) and the total molar flow rates of vapor and liquid (Figure 5).

Depropanizer

The bottom stream of the Deethanizer serves as feed stream to the Depropanizer. The overhead product of this column is propane rich and is condensed in the condenser. The reflux ratio and the mol% of propane in the distillate have been set equal to 5 and 85.0, respectively. Figure 6 shows the steady-state composition profile of this distillation column. The distillate stream has a flow rate equal to 30.58 kmole/hr and contains: 13.64 mol% ethane, 85 mol% propane and 1.33 mol% i-butane. The bottom stream has a flow rate equal to 36.07 kmole/hr and is composed of 18.82 mol% i-butane, 41 mol% n-butane, 23.55 mol% n-pentane, 11.08 mol% n-hexane and 5.55 mol% C_7^+. The cooling and heating duties of the condenser and reboiler are $2.429 \ 10^6$ and $2.348 \ 10^6$ kJ/hr, respectively. Again, perfect agreement was obtained with Aspen HYSYS$^®$, just as in the case of the Deethanizer. For simplicity, only compositions profiles comparison (continuous lines for Mathematica$^©$ and markers for HYSYS results) are shown in Figure 6.

Debutanizer

The stream exiting the Depropanizer reboiler serves as feed stream, after reduction of its pressure, to the Debutanizer. The reflux ratio and the mol% of n-butane in the residue have been set to 5 and 0.1, respectively. In Figure 7 the steady-state composition profiles of all constituents inside this distillation column are represented. i-butane and n-butane are separated out as overhead product. The distillate stream has a flow rate equal to 21.56 kmole/hr and contains a butane mixture with 31.48 mol% i-butane and 68.51 mol% n-butane. The bottom stream with a flow rate of 36.07 kmole/hr is composed of 58.56 mol% n-pentane, 27.55 mol% n-hexane and 13.87 mol% C_7^+. The cooling and heating duties of the condenser and reboiler are found to be $2.200 \ 10^6$ and $2.070 \ 10^6$ kJ/hr, respectively. Once more, perfect agreement was obtained with Aspen HYSYS$^®$ results.

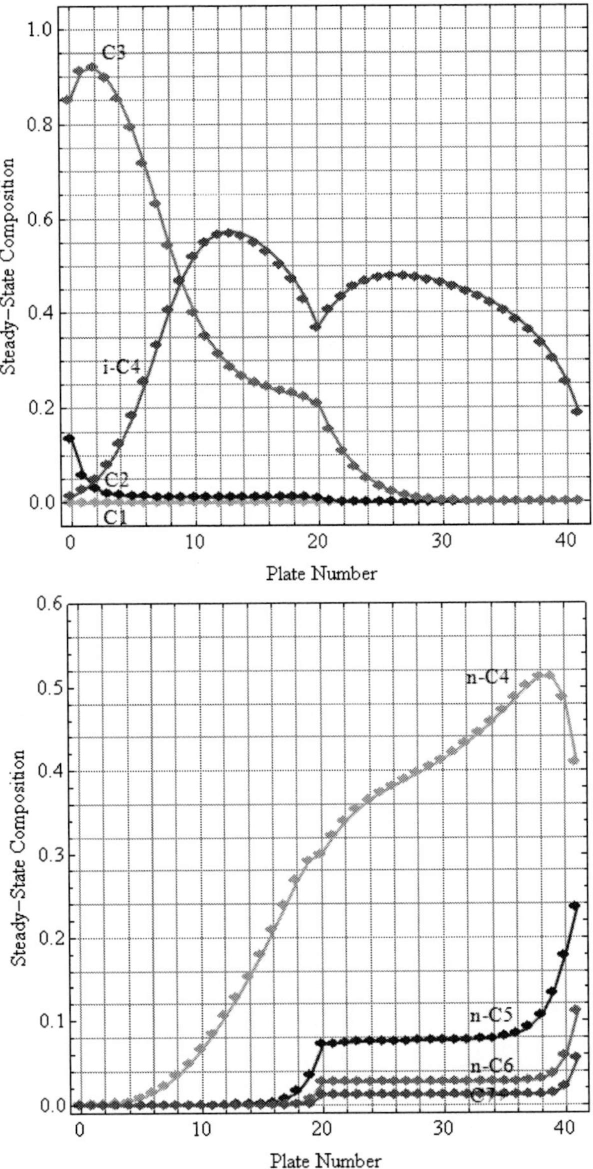

Figure 6. Steady-state composition profiles for the Depropanizer. (Continuous lines: Mathematica© and markers: Aspen HYSYS®).

Deisobutanizer

The liquid butanes mixture coming from the condenser of the preceding column will be separated now in its constituents n-butane and i-butane in the last distillate column of the NGL train, the Deisobutanizer. Figure 8 shows the composition evolution of i-butane and n-butane in the column. The distillate stream (flow rate: 6.70 kmole/hr) is composed of 98.98 mol% i-butane and 1.01 mol% n-butane and the bottom stream (14.86 kmole/hr), 1 mol% i-butane and 98.99 mol% n-butane. The cooling and heating duties of the condenser and reboiler are: 3.560 10^6 and 3.461 10^6 kJ/hr, respectively. We note again the almost perfect concordance between Aspen HYSYS[®] and Mathematica[©] simulation results.

For pedagogical purposes only, two more simulations are realized to illustrate the use of Mathematica[©] as tool for chemical engineering modeling and calculations.

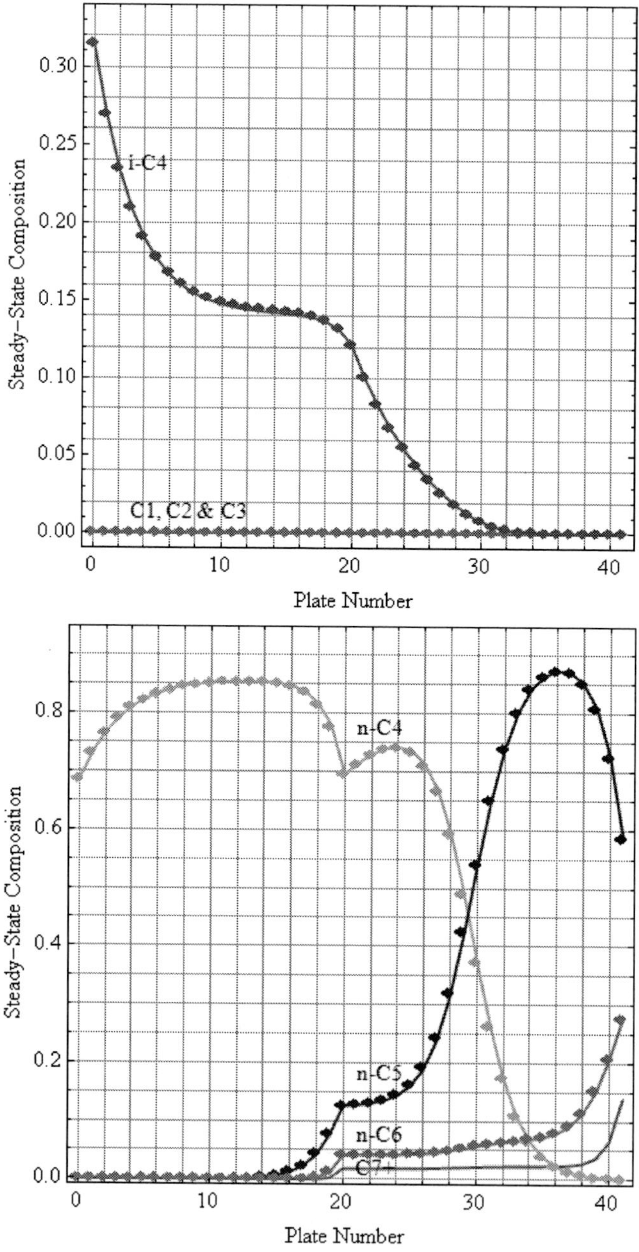

Figure 7. Steady-state composition profiles for the Debutanizer. (Continuous lines: Mathematica[©] and markers: Aspen HYSYS[®]).

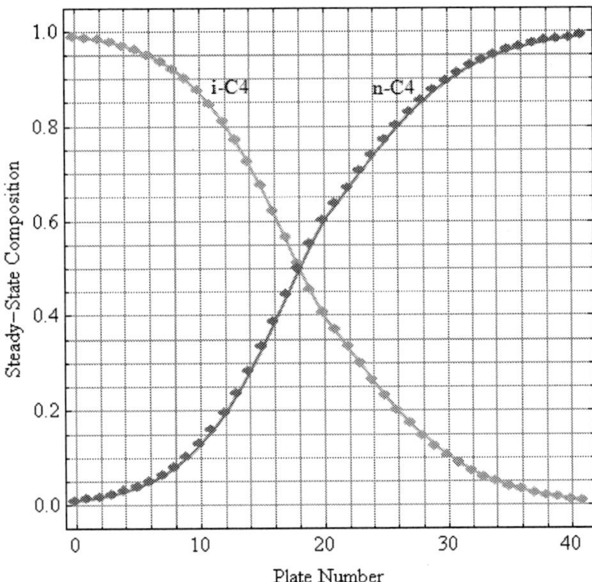

Figure 8. Steady-state composition profiles in the Deisobutanizer. (Continuous lines: Mathematica[©] and markers: Aspen HYSYS[®]).

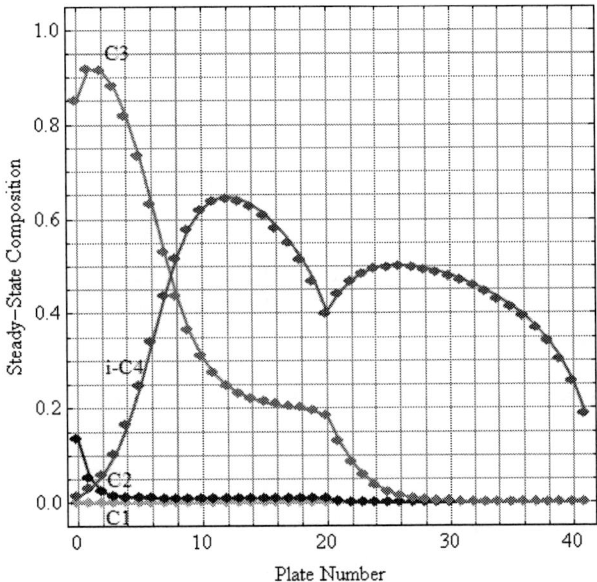

Figure 9. Steady-state composition profiles in the Depropanizer: case with pressure drop. (Continuous lines: Mathematica[©] and markers: Aspen HYSYS[®]).

Depropanizer with Pressure-Drop

For this case, we return to the Depropanizer but consider now the pressure drop in the column that was neglected previously. We set a Δp of 10 kPa for each stage and 20 kPa for condenser and reboiler. The reflux ratio and the mol% of propane in the distillate are set to 5 and 85.0, respectively, the same as before. Figure 9 shows the steady-state composition profiles of the lighter constituents in the distillation column (i.e., the Depropanizer), and Figure 10 the pressure evolution. No effect of the pressure drop on the distillate (13.61 mol% ethane, 85 mol% propane, and 1.37 mol% i-butane) and bottoms composition (18.79 mol% i-butane, 41.01 mol% n-butane, 23.56 mol% n-pentane, 11.08 mol% n-hexane and 5.56 mol% C_7^*) is observed. But we do now note an impact of the cooling and heating duties of the condenser and reboiler: 2.665 10^6 and 2.533 10^6 kJ/hr, respectively. More heat is supplied to the reboiler and more heat is rejected in the condenser, and hence more energy is consumed for the same separation by dissipative effects.

We notice once more the complete agreement with Aspen HYSYS® simulation results.

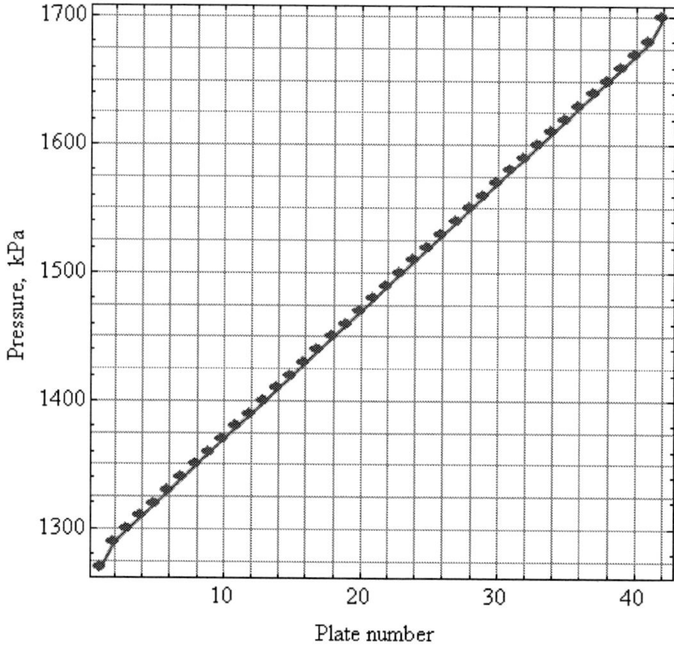

Figure 10. Pressure profile in the Depropanizer: case with pressure drop. (Continuous lines: Mathematica© and markers: Aspen HYSYS®).

Deisobutanizer with 75 % Liquid Murphree Efficiency

Real column plates rarely have the same performances as theoretical stages. Very often equilibrium is not reached and the separation is less efficient. To characterize real plate performances in comparison to theoretical separation under the same conditions, the Murphree efficiency is introduced [5]. It can be defined for the gas and for the liquid phase and for every tray in the column. To simplify the design and calculation of columns however, a mean Murphree efficiency is used for each phase.

We reconsider now the Deisobutanizer and assume a mean liquid Murphree efficiency of 75 % for otherwise similar operating conditions. In particular, the reflux and reboil ratios are set to 10 and 4.

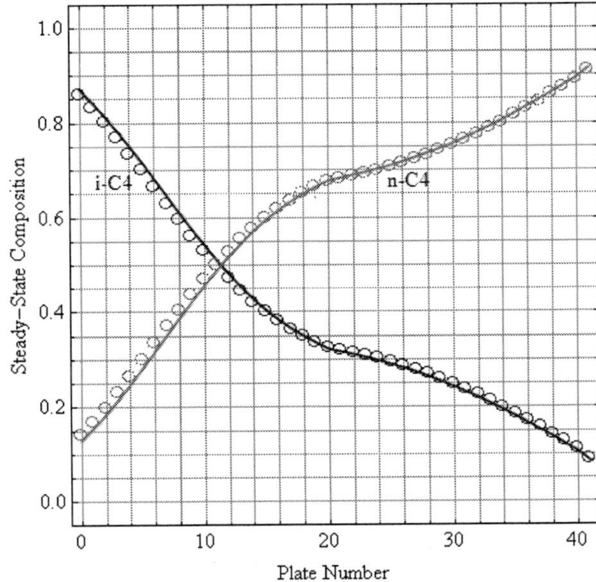

Figure 11. Steady-state composition profiles in the Deisobutanizer for 75% Murphree efficiency. (Continuous lines: Mathematica$^{©}$ and markers: Aspen HYSYS$^{®}$).

The effect of the separation process is rather important: a lower distillate stream flow rate is produced (6.291 kmole/hr) and the desired purity of the products are not reached: 87.02 mol% for *i*-butane and 91.46 mol% for *n*-butane. The cooling and heating duties of the condenser and reboiler are now 1.290 10^6 and 1.190 10^6 kJ/hr, respectively.

We observe again a complete concordance with Aspen HYSYS$^{®}$ results (Figure 11).

EXTRACTIVE DISTILLATION

Extractive distillation is a ubiquitous technique used for the separation of close-boiling mixtures as well as non-ideal binary mixtures presenting either a minimum-boiling or a maximum-boiling azeotrope [6]. Total annualized cost (TAC) calculations show that this method is more economical than heterogeneous azeotropic distillation [7]. In addition, this technique applies even if the azeotrope is not pressure-sensitive.

For low relative volatility separations a large number of stages and a high reflux ratio are necessary. Energy savings become then an important issue. Some solvents, called entrainer in the context of extractive distillation, alter selectively the relative volatility of one component when added to the mixture. This species is dissolved in the entrainer and exit the column as the bottom liquid mixture because the entrainer is normally the highest boiling product. This separation scheme necessitates generally two columns, one for extraction and one for solvent recovery (Figure 12).

In most industrialized countries, butadiene, an important monomer used in the production of synthetic rubber, is produced as a byproduct of ethylene and other olefin production processes by steam cracking [8]. If mixed with steam and subjected to pyrolysis, aliphatic hydrocarbons give up hydrogen to yield mixtures of unsaturated hydrocarbons, including butadiene. The quantity of butadiene produced depends on the nature of the aliphatic hydrocarbons fed to the steam cracker. Ethane cracking produces mainly ethylene. Heavier feeds favor the formation of higher molecular weight olefins, butadiene, and aromatic products. Butadiene is typically isolated from the other C_4 compounds produced by steam cracking using extractive distillation [8]. A polar aprotic entrainer such as acetonitrile, n-methyl pyrrolidone (NMP) or furfural is generally used as extractive solvent. This latter is then stripped from 1,3-butadiene by distillation.

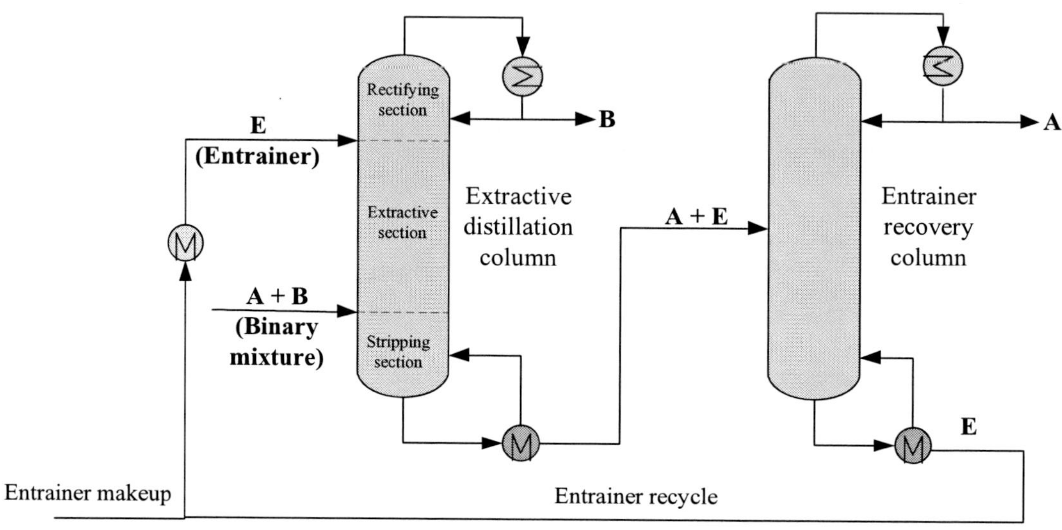

Figure 12. Set-up of the extractive distillation and regeneration columns.

As example of extractive distillation we consider a mixture composed of four close-boiling C_4 hydrocarbons: n-butane (normal boiling temperature: 272.7 K), i-butane (261.4 K), i-butene (266.2 K) and 1,3-butadiene (268.7 K) [9]. The objective is to separate 1,3-butadiene. The selected entrainer for our case study is furfural. As shown in Table 3, the relative volatilities of n-butane, i-butene and i-butane relative to 1,3-butadiene are close to unity and become larger if furfural is added to the mixture. One peculiar phenomenon is the fact that n-butane becomes more volatile than 1,3-butadiene in the presence of this solvent. Thus, the entrainer makes the separation of this close-boiling C_4 mixture possible simply by modifying the volatilities. It was shown that this effect can be further improved (i.e., much different from unity) if a small percentage of water is added to furfural [10].

Table 3. Relative volatilities of n-butane, i-butene and i-butane with respect to 1,3-butadiene at 448.15 kPa and 54.5 °C in the original mixture and after addition of furfural

	Original Mixture	Furfural Added
n-butane	0.8714	1.0777
i-butene	1.0449	1.2049
i-butane	1.1706	1.6066

We now study several aspects related to this separation method by performing simulations using Mathematica[©], and Aspen HYSYS[®] for comparison purposes. The Soave Redlich-Kwong EoS (Appendix A) will be used and in both simulations. The solvent regeneration column is also included in the simulations in order to show that the entrainer recovery is possible.

Extractive Distillation Column

We consider an extractive distillation column with 50 equilibrium stages, a partial reboiler and a total condenser and operating at 300 kPa. This column has an upper feed composed of pure furfural (i.e., the entrainer) with a flow rate equal to 55 kmole/hr. It is introduced at 20 °C at stage 3 counting from the top. The extractive column has also a lower feed composed of 5 mol% *n*-butane, 15 mol% *i*-butene, 20 mol% *i*-butane and 60 mol% 1,3-butadiene. The lower feed enters at stage 40 counting from the top at the same temperature of 20 °C with a flow rate of 10 kmole/hr. In the simulations, the mol% of 1,3-butadiene in the distillate and of *i*-butene in the residue are set equal to 3 and 0.4, respectively.

The mathematical model of the separation sequence consists of MESH set of 678 nonlinear algebraic equations, which are solved in less than 33 seconds using Mathematica[©].

Figures 13 and 14 show the steady state composition and temperature profiles in the extractive distillation column. It is clear from Figure 13 that furfural is an appropriate entrainer for the extraction of 1,3-butadiene. The distillate is composed of *n*-butane, *i*-butene and *i*-butane with a small fraction of 1,3-butadiene (i.e., 3 mol%). The bottom stream has a flow rate equal to 61.55 kmole/hr and is composed of mainly 1,3-butadiene and furfural (89.35 mol-% and 9.57 mol%, respectively). A non-negligible amount of *n*-butane (i.e., 0.65 mole %) is found in the bottom stream since the relative volatility of *n*-butane relative to 1,3-butadiene is very close to unity (1.077, Table 3). The heat and cooling duties of the reboiler and condenser are found equal to $1.509 \ 10^6$ kJ/hr and $0.786 \ 10^6$ kJ/hr, respectively.

Here also we notice the rather good agreement between the simulation results of Aspen HYSYS[®] (see Figure 15 for the extractive/recovery distillation set-up) and Mathematica[©].

Solvent Recovery Column

This column serves to recover the main product 1,3-butadiene and regenerate the solvent for recycling. The separation is realized in a 20 stages distillation column operating at 200 kPa and equipped with a partial reboiler and a total condenser. It is supplied with the bottoms of the extractive distillation column as feed that is introduced at stage 10 counting from the top. The reflux and reboil ratios are set to 5 and 0.777, respectively. Furfural is obtained as a bottom product and 1,3-butadiene as overhead stream.

The mathematical model of the recovery column is constituted of 288 nonlinear MESH equations system which is readily solved with Mathematica[©].

Figure 16 gives the steady-state composition profile for the different constituents. The distillate stream has a flow rate equal to 6.55 kmole/hr and is composed of 90.01 mol% 1,3-butadiene. The bottom stream is pure furfural and with a flow rate of 55 kmole/hr. A minor make-up solvent stream (flow rate: 0.001 kmole/hr) is necessary to compensate for evaporation. The condenser and reboiler duties are evaluated to 0.863 10^6 and 1.813 10^6 kJ/hr, respectively.

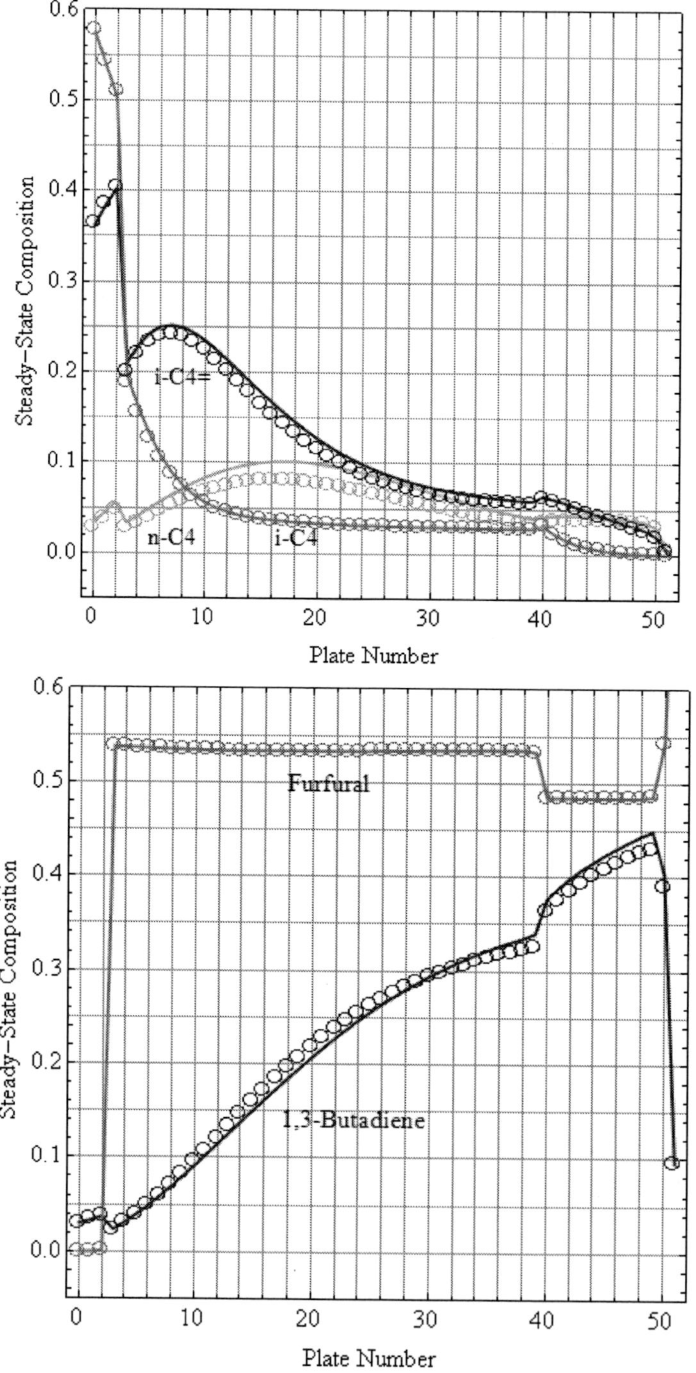

Figure 13. Steady-state composition profiles in the extractive distillation column. (Continuous lines: Mathematica© and markers: Aspen HYSYS®).

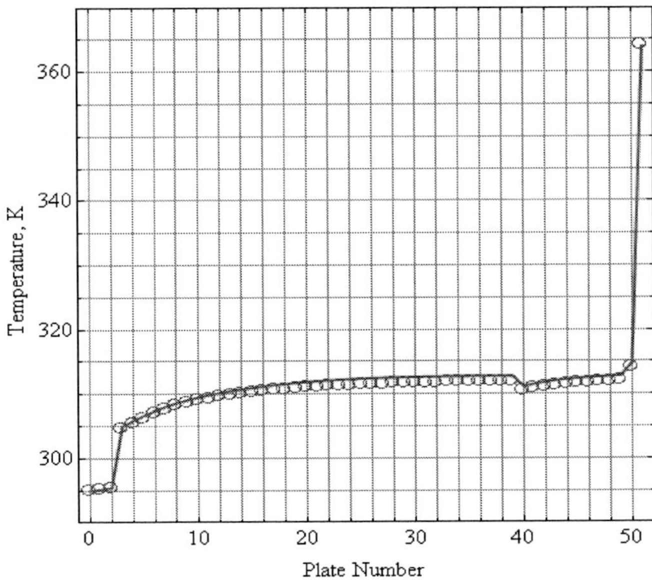

Figure 14. Temperature profile in the extractive distillation column. (Continuous lines: Mathematica[©] and markers: Aspen HYSYS[®]).

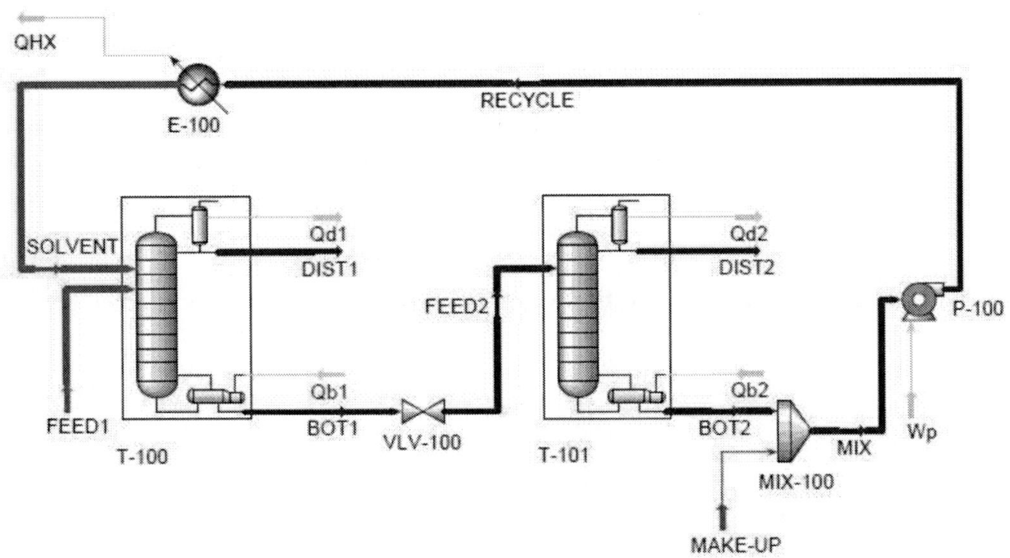

Figure 15. Aspen HYSYS[®] PDF for 1,3-butadiene furfural extraction.

REACTIVE DISTILLATION

In production processes involving chemical transformations, reaction and separation are usually handled in distinct devices. A simpler design integrating reaction and separation by distillation or distillation/liquid extraction in the same unit is in many instances possible and realized in one piece of equipment, a reactive distillation setup. The first applications of this technology were the alkylation of aromatics [11, 12, 13].

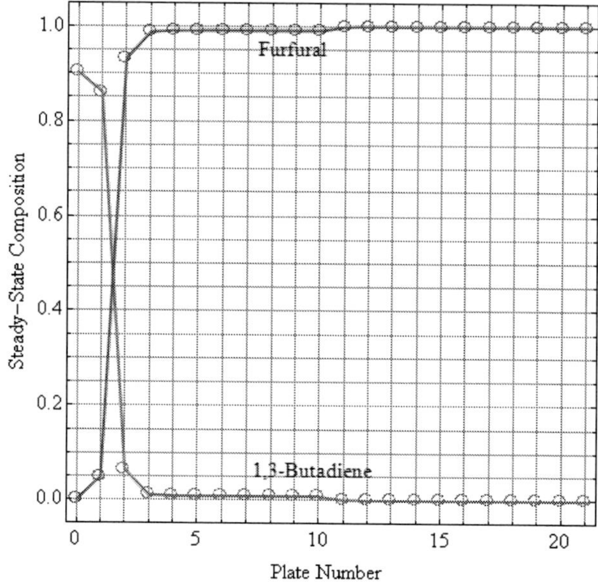

Figure 16. Steady-state liquid and vapor profiles in the extractive distillation column. (Continuous lines: Mathematica© and markers: Aspen HYSYS®).

With the continuous removal of reaction products, reactive distillation, when possible, offers the advantages of higher yields, energy savings and reduced capital costs [14].

The standard configuration of a reactive distillation column includes a rectification section, a reaction section and a stripping section. A set of reactive trays or reactive packing is used as reaction section.

Three applications of this innovative technology will be handled in the following.

Production of MTBE in a Reactive Distillation Column

Methyl-*t*-butyl ether, MTBE, is used as gasoline additive to increase its octane rating. It also allows the use of smaller amounts of lead alkyl compounds as anti-knock agents. It is further found that with this additive, the emission of carbon monoxide and nitrous oxides on combustion of gasoline is largely reduced [15]. The production reaction of methyl-*t*-butyl ether from *i*-butene and methanol takes place in the liquid phase in the presence of acidic ion-exchange resin as catalyst in a fixed bed reactor [16]. A simplified flow sheet of the process is represented in Figure 17.

Figure 17. (Continued).

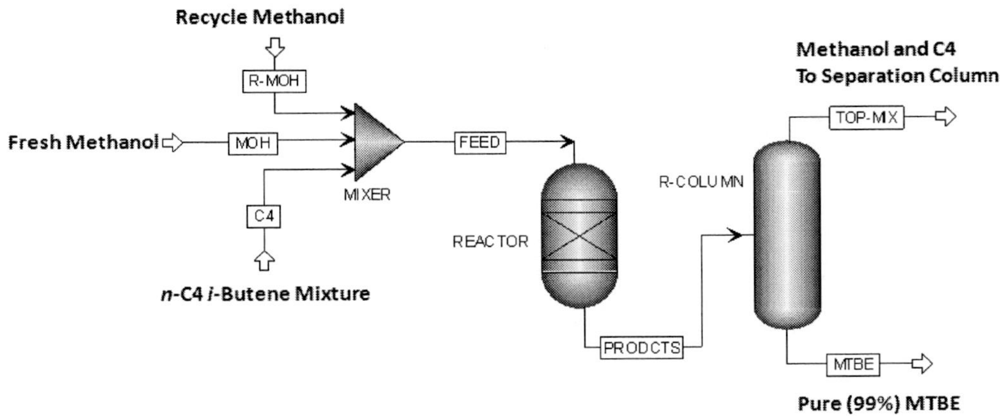

Figure 17. Schematics of the MTBE production process.

This case study deals with the simulation of the catalytic synthesis of methyl *tert*-butyl ether using reactive distillation. The objective is to achieve a high purity product, 99 % or better.

As a numerical example [17] we consider a reactive distillation column with 16 stages, a partial reboiler (stage 17) and a total condenser (stage 0). The reactive section where the synthesis reaction is taking place ranges from stage 3 to 9. The feed is a liquid mixture of 66.67% mole of *n*-butane as inert solvent and 33.33% mole of *i*-butene on one side, and pure methanol on the other side. It is introduced at stage 9. The operation pressure is set to 11.2 bars.

For the determination of the thermodynamic properties of the fluids and the calculation of the liquid-vapor equilibria the Wilson model [9, 18, 19, 20, 21] is used for the liquid phase and the Peng-Robinson equation of state (Appendix A) for the vapor phase.

The details of the Mathematica[©] model of the process can be found in reference [22]. We just emphasize here some results of this interactive program. As expected, no MTBE is formed for very low Damköhler numbers (molar hold-up near zero) and the only species exiting the column are the reactants *n*-butane, *i*-butene, and methanol.

At moderate Damköhler numbers, low conversions are achieved.

Finally, at high Damköhler numbers, almost pure MTBE (99% purity) is obtained as a bottom product and a mixture mainly composed of *n*-butane (88 mol%) as distillate (Figure 18). This composition is close to the binary nonreactive azeotrope mixture of methanol and *n*-butane.

As an exercise, the reader can run the Mathematica program to examine the effect of the molar hold up on the performances of the reactive distillation column.

Figure 19 presents a simplified flow sheet of the production of MTBE in a reactive distillation column as simulated with Aspen HYSYS[®] with similar operating conditions. As can be noted by comparing Figures 18 and 19, the simulation results obtained using Mathematica are in good concordance with those of Aspen HYSYS[®]. In particular, the bottom product meets the purity requirement imposed.

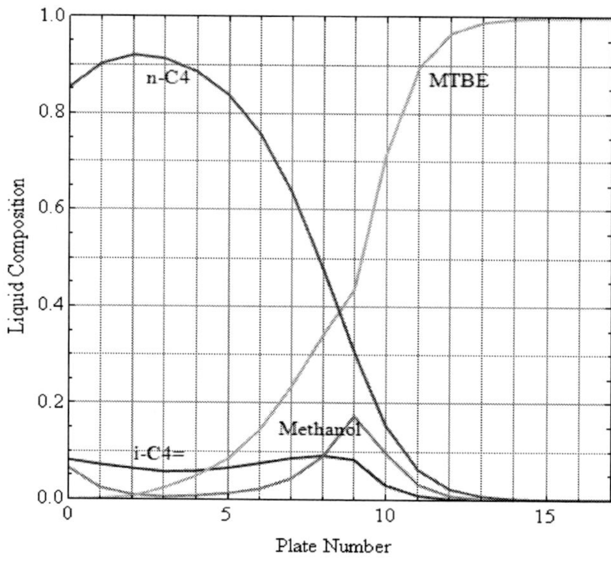

Figure 18. Profile of the compositions in the MTBE synthesis column.

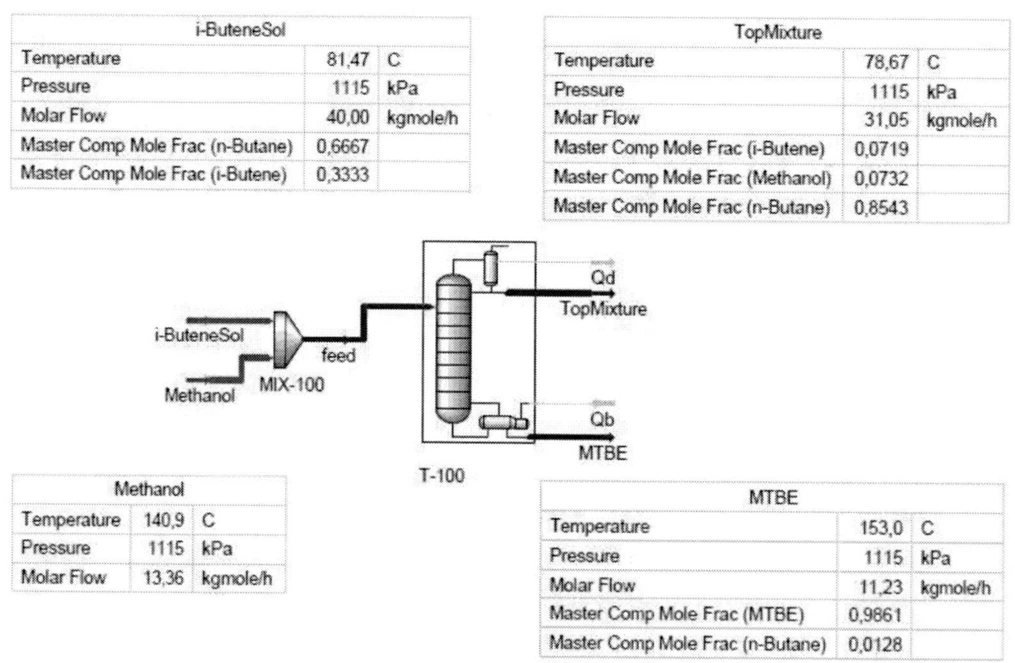

i-ButeneSol		
Temperature	81,47	C
Pressure	1115	kPa
Molar Flow	40,00	kgmole/h
Master Comp Mole Frac (n-Butane)	0,6667	
Master Comp Mole Frac (i-Butene)	0,3333	

TopMixture		
Temperature	78,67	C
Pressure	1115	kPa
Molar Flow	31,05	kgmole/h
Master Comp Mole Frac (i-Butene)	0,0719	
Master Comp Mole Frac (Methanol)	0,0732	
Master Comp Mole Frac (n-Butane)	0,8543	

Methanol		
Temperature	140,9	C
Pressure	1115	kPa
Molar Flow	13,36	kgmole/h

MTBE		
Temperature	153,0	C
Pressure	1115	kPa
Molar Flow	11,23	kgmole/h
Master Comp Mole Frac (MTBE)	0,9861	
Master Comp Mole Frac (n-Butane)	0,0128	

Figure 19. Aspen HYSYS® PDF of the reactive distillation column for the production of MTBE.

Decomposition of MTBE in a Reactive Distillation Column

Besides its major application as gasoline additive, MTBE offers also a method of selectively removing *i*-butene from a C$_4$-mixture. Its subsequent decomposition by cracking yields highly pure *i*-butene and methanol [16, 23].

In the present case study we simulate a reactive distillation column that produces *i*-butene and methanol from the decomposition of methyl *tert*-butyl ether (MTBE), the reverse reaction of that of the preceding case.

The reaction is taking place in the liquid phase in the presence of acid catalyst. Distilling the decomposition products under reflux yields an overhead fraction composed mainly of *i*-butene and bottoms effluent containing almost pure methanol.

The sieve tray column has 16 stages, a partial reboiler, and a total condenser. The column is fed with pure MTBE at stage 8. The reactive stages range from 6 to 11.

An asymmetric thermodynamic approach (ϕ–γ) is used to predict the properties of the fluid mixture and the calculation of the liquid-vapor equilibria: Wilson model for the liquid phase and ideal gas law for the vapor phase. The operation pressure is set to 11.2 bars. The details of the kinetics of this equilibrium limited reaction can be found in references [24, 25].

The Mathematica[©] interactive simulation model of this reactive distillation column can be found in reference [25]. Figures 20 and 21 are generated with this program. The reader is encouraged to run the program in order to study the effect of the operating conditions on the purity of the distillate.

For low Damköhler numbers (low molar holdup), the conversion is low. Figure 20 shows the composition profiles *vs.* stage number for the case of a molar holdup of 20 kmol. As can be noted unconverted MBTE is found in the bottoms mixed with the product methanol.

At high Damköhler numbers, we get almost pure methanol as bottoms and pure *i*-butene in the distillate (Figure 21). As Figure 22 depicts, similar results are obtained using Aspen HYSYS[®].

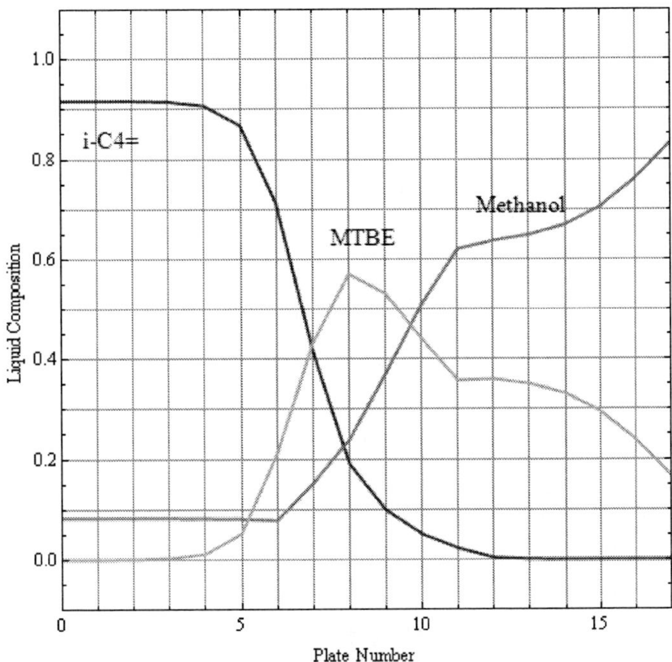

Figure 20. MTBE decomposition: Profiles of the compositions in the column in the case of a molar holdup of 20 kmol.

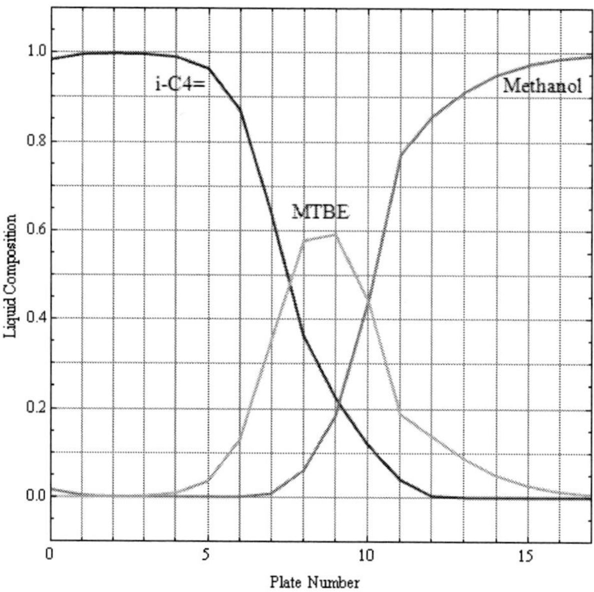

Figure 21. MTBE decomposition: Compositions in the for a molar holdup of 500 kmol.

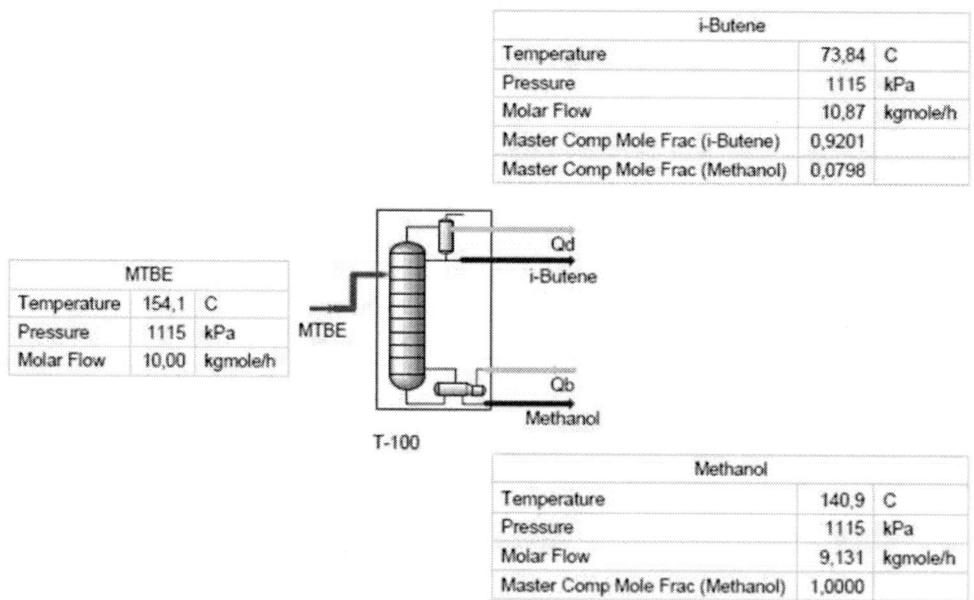

i-Butene		
Temperature	73,84	C
Pressure	1115	kPa
Molar Flow	10,87	kgmole/h
Master Comp Mole Frac (i-Butene)	0,9201	
Master Comp Mole Frac (Methanol)	0,0798	

MTBE		
Temperature	154,1	C
Pressure	1115	kPa
Molar Flow	10,00	kgmole/h

Methanol		
Temperature	140,9	C
Pressure	1115	kPa
Molar Flow	9,131	kgmole/h
Master Comp Mole Frac (Methanol)	1,0000	

Figure 22. Aspen HYSYS® model of the reactive distillation column for the decomposition of MTBE (Reflux ratio: 7.5; Boilup ratio: 8.25).

Metathesis of cis-2-pentene in a Reactive Distillation Column

We consider at last case the disproportionation reaction, known also as a metathesis, of cis-2-pentene to cis-2-hexene and cis-2-butene.

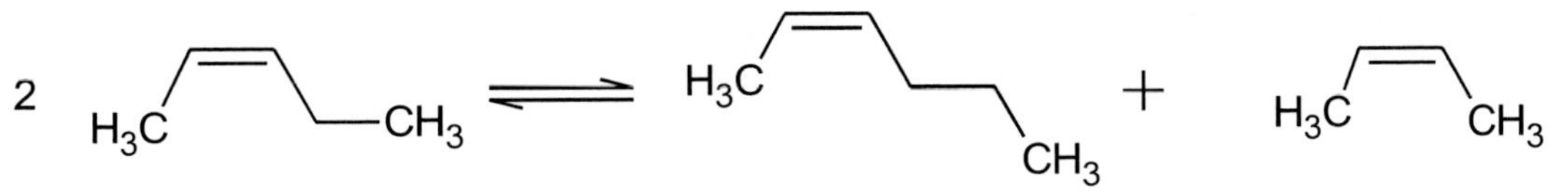

In order to simplicity the notations, we designate by C, B and A respectively the components cis-2-pentene, cis-2-hexene and cis-2-butene.

The ternary mixture is subject to an equilibrium-limited chemical reaction with reaction rate [26, 27]

$$r = 0.5\,k\left[x_C^2 - \frac{x_A\,x_B}{K_{eq}} \right]$$

where K_{eq} is the temperature dependant equilibrium constant, and

$$k = 3553.6\exp\left[-\frac{6000}{RT} \right]$$

the reaction rate constant, with $R = 1.987$ cal mol^{-1} K^{-1}.

The pure reactant C (cis-2-pentene) is fed to a reactive distillation column operating at a pressure of 3 atmospheres with 13 plates; the feed stage location is stage 5, the reactive stages go from stages 2 to 7. The feed flow rate is set to 100kmol/hr.

The following simulations are made with the simplifying assumption of constant molar overflow (CMO) and by neglecting heat effects [6, 27, 28].

The Mathematica$^{©}$ interactive simulation model of this reactive distillation column can be found in reference [28]. Figures 23, 24 and 25 illustrate some results of the calculations, namely:

- the composition profile *vs.* plate number for the three components of the ternary system (A in red, B in blue, C in green and the reactive zone in light blue) (Figure 23),
- the temperature profile in the column (Figure 24), and
- the ternary diagram with the composition (mole fraction) of B *vs.* the composition of A. The feed composition is shown by a dot and the reactive stages 2 to 7 are displayed (Fig. 25).

It is worth noting that for high hold-up values (i.e., large Damköhler numbers) one recovers the equilibrium case shown by the green solid curve in the ternary diagram.

We note again the good concordance between the simulations results of the Mathematica program with those obtained with the Aspen HYSYS$^{®}$ software (Figure 26).

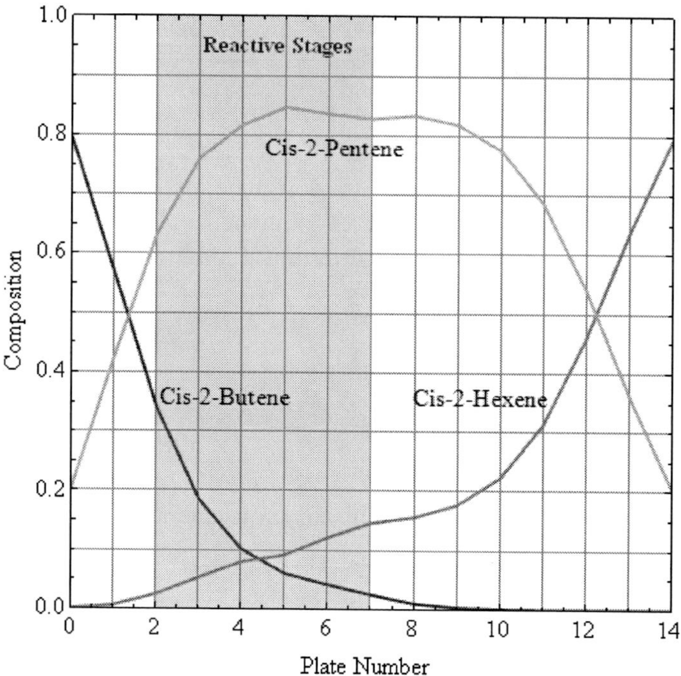

Figure 23. Compositions in the metathesis column (Reflux ratio: 14; Molar holdup: 500 kmol).

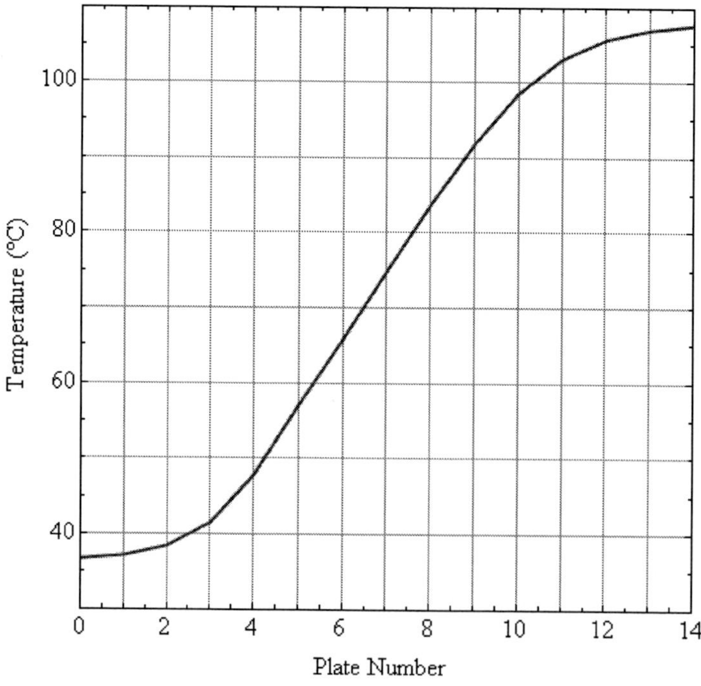

Figure 24. Profile of temperature in the metathesis column (Reflux ratio: 14; Molar holdup: 500 kmol).

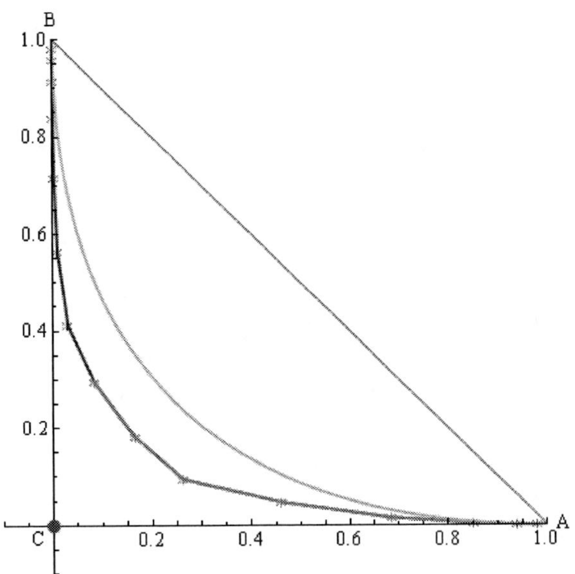

Figure 25. Tarnary diagram (Equilibrium curve in green) (Reflux ratio: 6.5; Molar holdup: 450 kmol).

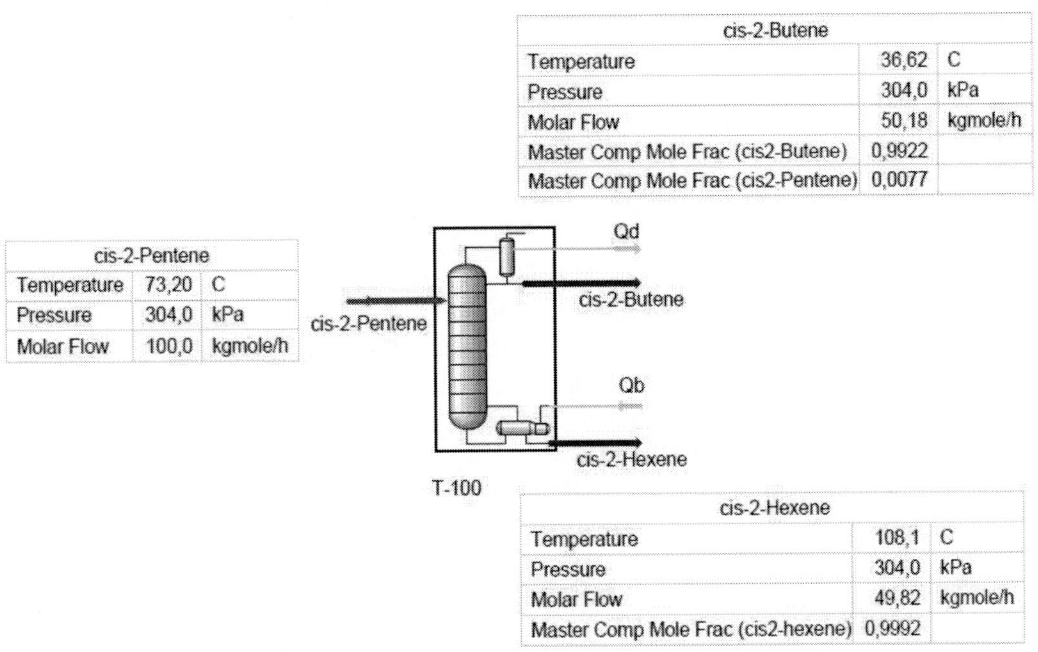

cis-2-Butene		
Temperature	36,62	C
Pressure	304,0	kPa
Molar Flow	50,18	kgmole/h
Master Comp Mole Frac (cis2-Butene)	0,9922	
Master Comp Mole Frac (cis2-Pentene)	0,0077	

cis-2-Pentene		
Temperature	73,20	C
Pressure	304,0	kPa
Molar Flow	100,0	kgmole/h

cis-2-Hexene		
Temperature	108,1	C
Pressure	304,0	kPa
Molar Flow	49,82	kgmole/h
Master Comp Mole Frac (cis2-hexene)	0,9992	

Figure 26. Aspen HYSYS® model of the reactive distillation column for the cis-2-pentene metathesis with operating conditions and product properties.

EXPLORING FURTHER PEDAGOGICAL PROBLEMS USING MATHEMATICA$^{©}$

In all the distillation simulations presented for the different case studies, no optimizations have been attempted. Indeed, the number of trays, the feed tray locations as well as the reflux and reboil ratios could be varied easily in order to come-up with more economical designs. For this purpose, TAC calculations [29] could be performed. In addition, column diameters could be determined based on the classical procedure by Fair [30, 31]. In principle, it would also be possible with the help of Mathematica$^{©}$ to perform even dynamic simulations for these distillation columns [32] and then to device appropriate control schemes [33].

For the case of extractive distillation, solvents other than furfural could be used. For example, NMP is known to be a very good candidate for the purification of C_4 mixtures containing 1,3-butadiene. Such calculations would require: (1) a good activity coefficient prediction model such as the UNIQUAC model [9, 19, 20, 34] to allow for non-ideal behavior in the liquid phase and (2) an EoS such as the SRK or PR EoS to predict the gas-phase fugacity coefficients since the column operates at relatively high pressures.

CONCLUSION

In this chapter five industrially relevant fractionation processes by different distillation techniques are simulated: the separation of natural gas liquids (NGLs) in a distillation train, the extractive distillation of 1,3-butadiene from a C_4 cut with furfural as the entrainer, the production of MTBE from i-butene and methanol in a reactive distillation column, the reverse process of production of i-butene by the decomposition of methyl *tert*-butyl ether, and the metathesis of cis-2-pentene to cis-2-butene and cis-2-hexene.

Routinely these processes are treated readily using a flow sheeting software like Aspen-HYSYS$^{®}$, similarly to what is done here for comparison purposes. The objective of this chapter was to show how these complex processes can also be handled with more pedagogical benefits using the computer algebra Mathematica$^{®}$. In all investigated cases, the comparison between the two approaches of simulations shows very good agreement.

The Mathematica$^{©}$ interactive simulation models developed for the different case studies are available at http://demonstrations.wolfram.com.

REFERENCES

[1]   Devold H. *Oil and Gas Production Handbook*, Edition 2.3, Oslo: ABB, 2010.

[2]   Kidnay, A. J. & Parrish, W. R. *Fundamentals of Natural Gas Processing*, Boca Raton: Taylor and Francis, 2006.

[3]   Maddox R. N. & Erbar J. H. *Gas Conditioning and Processing*, Campbell Petroleum Series, Volume 3, Oklahoma: 1992.

[4]   Younger, A. H. & Eng., P. *Natural Gas Processing Principles and Technology*, Part II, University of Calgary, 2004.

[5] Wankat, P. C. *Separation Process Engineering*, 2nd Edition, Upper Saddle River: Prentice Hall, 2007.

[6] Doherty, M. F. & Malone, M. F. *Conceptual Design of Distillation Systems*, New York: McGraw–Hill, 2001.

[7] Lei, Z., Li, C. & Chen, B. "Extractive Distillation: a Review," *Separation and Purification Reviews*, 32, pp. 121-213, 2003.

[8] Sun, H. P. & Wristers, J. P. *Butadiene, Encyclopedia of Chemical Technology*, Volume 4, 4th Edition, New York: John Wiley & Sons, 1992.

[9] Reid, C. R., Prausnitz, J. M. & Poling, B. E. *The properties of Gases and Liquids*, 4th Edition, McGraw-Hill, Inc., 1987.

[10] Buell C. K., & Boatright R. G. "Furfural Extractive Distillation," *Industrial and Engineering Chemistry*, *39,* pp. 695–705, 1947.

[11] Smith, L. A. US Patent 4 849 569, 1989; US Patent, 5 446 223, 1995.

[12] Hsieh *et al.* US Patent 5 082990, 1992.

[13] Dimian, A. C. & Bildea, C. S. *Chemical Process Design*, Weinheim: Wiley-VCH Verlag GmbH & Co., 2008.

[14] *Process Modeling Using HYSYS With Chemical Industry Focus*, ASPEN HYSYS Documentation, 2004.

[15] Adams, J. M., Clement, D. E. & Graham, S. H. *Clays and Clay Minerals*, Vol. 30, No. 2, pp. 129-134, 1982.

[16] Speight, J.G *Chemical and Process Design Handbook*, McGraw-Hill, Inc., 2002.

[17] Chen, F., Huss, R. S., Malone, M. F. & Doherty, M. F. "Simulation of Kinetic Effects in Reactive Distillation" *Computers and Chemical Engineering*, 24(11), pp. 2457–2472, 2000.

[18] Wilson, G. M. "Vapor-Liquid Equilibrium XI: a new expression for the excess free energy of mixing", *Journal of the American Chemical Society*, Vol. 86, pp. 127-130, 1964.

[19] Sandler, S. I. *Chemical Engineering Thermodynamics*, 3rd Edition, John Wiley and Sons, 1999.

[20] Poling, B. E., Prausnitz, J. M. & O'Connell, J. P. *The properties of Gases and Liquids*, 5th Edition, McGraw-Hill, Inc., 2001.

[21] Binous, H. Computation of Residue Curves Using Mathematica and Matlab, in *Computer Simulations*, Boris Nemanjic and Navenka Svetozar Editors, Nova Science Publishers, Inc., pp. 161-173, 2013.

[22] Binous, H., Selmi, M., Wada, I., Allouche, S. & Bellagi, A. "Methyl *Tert*-Butyl Ether (MTBE) Synthesis with a Reactive Distillation Unit", Wolfram Demonstrations Project: http://demonstrations.wolfram.com/MethylTertButylEtherMTBESynthesisWithAReacti veDistillationUn/

[23] Deguchi T. & Tokumaru T. Patent EP 0068785A1, 1983.

[24] Huang K. & Wang S. J. "Design and Control of a Methyl Tertiary Butyl Ether (MTBE) Decomposition Reactive Distillation Column", *Ind. Eng. Chem. Res.,* 46(8), pp. 2508–2519, 2007.

[25] Binous H., Selmi M., Wada I. Allouche S. & Bellagi A. "Methyl Tert-Butyl Ether (MTBE) Decomposition with a Reactive Distillation Unit", Wolfram Demonstrations Project:

http://demonstrations.wolfram.com/MethylTertButylEtherMTBEDecompositionWithA
ReactiveDistillation/

[26] Dragomir, R. M. & Jobson, M. "Conceptual Design of Single-Feed Kinetically Controlled Reactive Distillation Columns", *Chemical Engineering Science,* 60(18), pp. 5049–5068, 2005.

[27] Doherty, M. F. & Knapp, J. P. "Distillation, Azeotropic and Extractive," in *Kirk-Othmer Encyclopedia of Chemical Technology*, New York: John Wiley & Sons, 2004.

[28] Binous, H., Selmi, M., Wada, I. Allouche, S. & Bellagi, A. "Production of Cis2-Butene and Cis2-Hexene by Cis2-Pentene Disproportionation", Wolfram Demonstrations Project, http://demonstrations.wolfram.com/ProductionOfCis2ButeneAndCis2 Hexene ByCis2PenteneDisproportion/

[29] Douglas, J. M. *Conceptual Design of Chemical Processes*, International Edition, New York: McGraw–Hill, 1988.

[30] Perry, R. H. & Green, D. W. Editors. *Perry's Chemical Engineers' Handbook*, International Edition, 7th Edition, New York: McGraw–Hill, pp. 14-27, 1997.

[31] Seader, J. D. & Henley, R. H. *Separation Process Principles*, New York: John Wiley and Sons, Inc., 1998.

[32] Nasri, Z. & Binous, H. "Rigorous Distillation Dynamics Simulations Using a Computer Algebra", *Computer Applications in Engineering Education*, 20, pp. 193–202, 2012.

[33] Luyben, W. L. *Distillation Design and Control Using Aspen Simulation*, New York: Wiley, 2006.

[34] Tester, J. W. & Modell, M. *Thermodynamics and its Applications,* 3rd Edition, Prentice Hall International, 1997.

[35] Peng, D. Y. & Robinson D. B. "A New Two-Constant Equation of State", *Industrial and Engineering Chemistry: Fundamentals,* 15, pp. 59-64, 1976.

[36] Soave, G. "Equilibrium Constants from a Modified Redlich-Kwong Equation of State", *Chemical Engineering Science*, 27, pp. 1197–1203, 1972.

[37] Nasri, Z. & Binous H. "Applications of the Soave-Redlich-Kwong Equation of State Using Mathematica®", *Journal of Chemical Engineering of Japan*, 40, pp. 534-538, 2007.

[38] Nasri, Z. & Binous H. "Applications of the Peng-Robinson Equation of State using MATLAB", *Chemical Engineering Education*, 43, pp. 115-124 (2009).

APPENDIX A:
CUBIC EQUATIONS OF STATE

Cubic equation of states are pressure explicit equation in temperature and volume that, when expanded, would contain the volume raised to the third power. They are often used in the oil and gas and petrochemical industries to describe the behavior of the fluid mixtures. They allow the prediction of several important aspects of chemical engineering thermodynamics such as vapor-liquid equilibrium data, liquid-phase and vapor-phase compressibility factors, enthalpies of gases and liquids, etc. Simulations of the distillation of several industrially relevant hydrocarbon mixtures commonly build on equations of states such as the SRK and PR EoS.

van der Waals Equation of State

This well-known equation from which the other cubic EoS are derived, writes

$$P = \frac{RT}{V-b} - \frac{a}{V^2} \tag{A-1}$$

where a and b are fluid specific constants that can be evaluated from the critical constants (pressure, P_c and temperature, T_c) of the substance [9],

$$a = \frac{27}{64}\frac{R^2 T_c^2}{P_c} \quad b = \frac{R\ T_c}{8 P_c} \tag{A-2}$$

Redlich-Kwong Equation of State

This modification of the van der Waaals EoS predicts better the vapor-liquid equilibrium of pure fluids. It writes

$$P = \frac{RT}{V-b} - \frac{a}{\sqrt{T}\,V(V+b)} \tag{A-3}$$

The fluid specific constants a and b are now expressed as [9, 19]

$$a = 0.42748\frac{R^2 T_c^{2.5}}{P_c} \quad b = 0.08664\frac{R\ T_c}{P_c} \tag{A-4}$$

Peng-Robinson Equation of State

This equation, as well as the Soave-Redlich-Kwong equation, is often used to describe the volumetric behavior of fluid mixtures in the oil and gas industry and to deduce the relevant thermodynamic properties as illustrated in the case of the SRK EoS. The expression of the pressure is now [9, 19, 34, 35]

$$P = \frac{RT}{V-b} - \frac{a}{V(V+b)+b(V-b)} \tag{A-5}$$

where the constants a and b are given as follows [9, 19]:

$$a = 0.45724 \frac{(RT_c)^2}{P_c}\left(1 + m\left(1 - \sqrt{T_r}\right)\right)^2 \qquad \text{(A-6)}$$

$$b = 0.07780 \frac{RT_c}{P_c} \qquad \text{(A-7)}$$

$$T_r = \frac{T}{T_c} \qquad \text{(A-8)}$$

$$m = 0.37464 + 1.5422\,\omega - 0.26992\omega^2 \qquad \text{(A-9)}$$

ω is the Pitzer acentric factor, also a fluid specific constant [9].

Soave-Redlich-Kwong Equation of State

For a pure component, the SRK EoS [9, 19, 36, 37] is given by,

$$P = \frac{RT}{V - b} - \frac{a}{V(V + b)} \qquad \text{(A-10)}$$

where the constant a is set as a function of the temperature

$$a = 0.42748 \frac{(RT_c)^2}{P_c}\left(1 + m\left(1 - \sqrt{T_r}\right)\right)^2 \qquad \text{(A-11)}$$

$$b = 0.08664 \frac{RT_c}{P_c} \qquad \text{(A-12)}$$

$$m = 0.480 + 1.574\,\omega - 0.176\,\omega^2 \qquad \text{(A-13)}$$

The cubic character of these EoS can be best seen when the equation is written in terms of the compressibility factor, Z ($Z = PV/RT$)). For the SRK EoS we get an equation in the form

$$Z^3 - \alpha Z^2 + \beta Z - \gamma = 0 \qquad \text{(A-14)}$$

For a fluid mixture with C components, this equation writes

$$Z^3 - Z^2 + Z\left(A - B - B^2\right) - AB = 0 \qquad \text{(A-15)}$$

where A and B are deduced from the individual component constants [19]

$$A = \sum_{i=1}^{C}\sum_{j=1}^{C} y_i y_j A_{ij} \quad \text{or} \quad \sum_{i=1}^{C}\sum_{j=1}^{C} x_i x_j A_{ij} \qquad \text{(A-16)}$$

$$A_{ij} = \sqrt{A_i A_j}\,(1 - k_{ij}) \qquad \text{(A-17)}$$

$$B = \sum_{i=1}^{C} y_i B_i \quad \text{or} \quad \sum_{i=1}^{C} x_i B_i \qquad \text{(A-18)}$$

$$A_i = 0.42747 a_i \frac{P_{r_i}}{T_{r_i}^2} \quad \text{and} \quad B_i = 0.08664 \frac{P_{r_i}}{T_{r_i}} \qquad \text{(A-19)}$$

The reduced pressure and temperature for each component are given by $P_r = P/P_c$ and $T_r = T/T_c$ and the parameter a_i for the pure component i is given by the equation (A-11).

For the Mathematica$^{©}$ simulations, the values of the binary interaction parameters (i.e., k_{ij}) are all taken from the Aspen HYSYS$^{®}$ data bank.

The equilibrium constants are obtained using the (ϕ–ϕ) method as follows [19],

$$K_i = \frac{\phi_{l_i}}{\phi_{v_i}} \quad \text{for } i = 1 \text{ to } C \qquad \text{(A-20)}$$

where the vapor fugacity coefficient is given by

$$\phi_{v_i} = \exp\left((Z_{v_i} - 1)\frac{B_i}{B} - \ln(Z_{v_i} - B) - \frac{A}{B}\left(\frac{2 A_i^{0.5}}{A^{0.5}} - \frac{B_i}{B} \right)\ln\left(\frac{Z_{v_i} + B}{Z_{v_i}} \right) \right) \qquad \text{(A-21)}$$

A similar expression is obtained for the liquid phase fugacity coefficient, ϕ_{l_i}, by replacing the gas-phase compressibility factor, with its liquid-phase counterpart, Z_{l_i}.

The liquid and vapor mole fractions are related by

$$y_i = K_i x_i \text{ with } i = 1 \text{ to } C \qquad \text{(A-22)}$$

In addition, the departure function from ideality for the enthalpy is given by [19],

$$H^{D} = RT(Z-1) + \cfrac{1}{B\cfrac{RT}{P}} Log\left(\frac{Z+B}{Z}\right)\left[T\frac{d}{dT}\left(A\frac{(RT)^2}{P}\right) - A\frac{(RT)^2}{P}\right] \qquad \text{(A-23)}$$

Similar equations are available for the PR EoS [19, 38],

APPENDIX B
MASS AND ENTHALPY BALANCE (MESH) EQUATIONS

F, L, and V are the molar flow rates of the feed stream, the liquid and the vapor (kmol/hr), respectively. x designates the mole fraction in the liquid phase, and y, in the vapor phase.

- Feed tray (k = f) balances

$$F + L_{f-1} + V_{f+1} - L_f - V_f = 0 \qquad \text{(B-1)}$$

$$F z_{f,i} + L_{f-1} x_{f-1,i} + V_{f+1} y_{f+1,i} - L_f x_{f,i} - V_f y_{f,i} = 0 \quad (i = 1,2,\ldots,C) \qquad \text{(B-2)}$$

$$F h_F + L_{f-1} h_{f-1} + V_{f+1} H_{f+1} - L_f h_f - V_f H_f = 0 \qquad \text{(B-3)}$$

- MESH equations for the kth tray (k $\neq$ f and $1 \leq k \leq N$)

$$L_{k-1} + V_{k+1} - L_k - V_k = 0 \qquad \text{(B-4)}$$

$$L_{k-1} x_{k-1,i} + V_{k+1} y_{k+1,i} - L_k x_{k,i} - V_k y_{k,i} = 0 \quad (i = 1,2,\ldots,C) \qquad \text{(B-5)}$$

$$L_{k-1} h_{k-1} + V_{k+1} H_{k+1} - L_k h_k - V_k H_k = 0 \quad (i = 1,2,\ldots,C) \qquad \text{(B-6)}$$

- Reflux drum (k = 0) balances

$$V_1 - (L_0 + D) = 0 \qquad \text{(B-7)}$$

$$V_1 y_{1,i} - (L_0 + D) x_{D,i} = 0 \quad (i = 1,2,\ldots,C) \qquad \text{(B-8)}$$

$$V_1 H_1 - (L_0 + D) h_0 - \dot{Q}_C = 0 \quad (i = 1,2,\ldots,C) \qquad \text{(B-9)}$$

- Reboiler (k = N+1) balances

$$L_N - B - V_{N+1} = 0 \qquad \text{(B-10)}$$

$$L_N x_{N,i} - B x_{N+1,i} - V_{N+1} y_{N+1,i} = 0 \quad (i = 1,2,\ldots,C) \qquad \text{(B-11)}$$

$$L_N h_N - B h_{N+1} - V_{N+1} H_{N+1} + \dot{Q}_B = 0 \quad (i = 1,2,\ldots,C) \qquad \text{(B-12)}$$

In: Advances in Systems Engineering Research
Editors: Elena Fermi and Adam Lamberti

ISBN: 978-1-62948-310-8
© 2013 Nova Science Publishers, Inc.

Chapter 4

ENTERPRISE TRANSFORMATION FROM THE GROUND UP: ADDRESSING INDIVIDUAL PERCEPTUAL AND BEHAVIORAL BIASES AS SCALING FRACTALS TO CREATE EMERGENT STATE CHANGES

Eric D. Smith[1] and Aditya Akundi[2]

[1]Systems Engineering Graduate Program Coordinator, Research Associate,
Research Institute for Manufacturing and Engineering Systems,
Industrial, Manufacturing and Systems Engineering Department,
University of Texas, El Paso, Texas, US
[2]Research Assistant, Research Institute of Manufacturing and Engineering Systems,
Doctoral Student, Electrical & Computer Engineering, Systems Engineering Track,
University of Texas, El Paso, TX, US

ABSTRACT

Organizations are complex adaptive systems with basic elements of individual persons, deciding and acting according to their own personal logic and behavior. Effective collective action is known to best occur when individuals are aligned toward a unifying corporate vision. Cognitive biases and heuristics, however, have been evolutionarily embedded into individual decision making so as to minimize individual risks, as exhibited through behavioral focus on short-term goals, bounded rationality, and the use of intuition as acquired through personal experience. The transition from a personal level of perception to a collective behavioral space and the contextual organizational setting is mediated by the Fundamental Attribution Error, the Correspondence Bias, and the Actor-Observer Bias at the cognitive level. The organizational effects of ameliorating and modifying individual biases are predicted through Balance Theory models that account for interpersonal behavioral changes that lead toward cognitive consistency among individuals and that can eventually lead to emergent and lasting enterprise transformations. The principal aim of this research is disrupt the traditional paradigm of dissipative managerial direction with demonstrable alternatives that enable common and beneficent personal decisions to rapidly scale up, creating holistic state changes in enterprises.

1. Organization As a Complex System with System Elements Being Individuals

1.1. Complex Adaptive Systems

A Complex System can be defined as a system which is composed of many independent system elements playing a key important role in the whole system's behavior. While the independent system elements follow their own logic and behavior, and interact among themselves, such elements define the dynamic behavior observed in a collectively Complex system. Based on the principles of Complexity Science, the complexity in any system can be characterized by understanding the relations, interactions and the behavior of the constituent system elements. According to Weaver (1948) systems can be formalized by two distinctions, Disorganized Complexity and Organized Complexity. In Weaver's view, disorganized complexity is a result of many constituent system elements. In such a case, the system elements interact among several others, contributing to random system behavior. To understand such a system, Statistical and Probabilistic methods can be used. On the other hand, organized complexity, in Weaver's view, is a non- random behavior of the system elements where the number of parts need not be large for the system to be knowledge emergent. Properties of such a system can be understood with the help of simulations and various modeling techniques. For example, we can observe organized complexity in Self-Sustainable cities and disorganized complexity in the behavior of planets and their orbital rotations.

A Complex Adaptive System (CAS) can be seen as a special type of a complex system, usually defined as a collection of several connected elements with an ability of adapting to their environment. A CAS can also be contemplated as a Dynamic Network Structure with an inherent ability to adapt, learn and behave accordingly, based on its environment. According to Holland (2006) a CAS can also be defined as a system with large number of components, often called agents or constituent system elements, that interact, adapt and learn. Figure 1, developed below, gives a high level illustration of the system elements and their knowledge emergence where $i_1.....i_n$ represent the system inputs and $o_1....o_n$ represent the system outputs charactering the system behavior.

1.1.1. Characteristics of a Complex Adaptive System

Complexity can be a result of the number of inter-relationships and interconnectivities among elements within a system and between its elements and the environment. As a result of this, a system can be characterized based on its behavior, which is prone to change and emerge as a result of spatial interactions among different levels. According to Chan (2001) a few attributes of complex adaptive systems are

- Distributed Control: No single control mechanism to govern a system's behavior
- Connectivity: Implies that an action taken by one part of a system will influence all the other parts of a system in either a linear or a nonlinear fashion
- Co-Evolution: Change of behavioral patterns over time based on the internal interactions as well as with the operating environment

- Sensitivity and Dependence: Unpredictable behavior over a period of time because a tiny disturbance can potentially affect the system harmonization and result in divergent behavior
- Emergent order: No explicit rule to define a system's behavior over a period of time; the system self-organizes itself based on knowledge emergence.

1.2. Organizations

An Organization can be defined as a collection of physical and social entities working together in order to achieve a common collective goal. There are several relations, interactions, and activities among the individuals involved that are necessary to effectively function towards a predefined enterprise vision. Assigned roles, responsibilities, tasks and authorities help to evaluate a situation or a project and take decisions collectively. A typical hierarchical structure in an organization helps to determine the roles and responsibilities assigned. According to Berkowitz and Wolff (1999), the 3 main elements of an organizational structure irrespective of the type of organization are: Governance *(in charge of taking decisions in an organization)*, organizational Operative Rules *(Rules established for proper functionality and to make sure the process and methodologies incline towards the whole enterprise vision)* and Work Distribution *(division of labor)*.

1.2.1. Characteristics of Organizations

Organizations are established and developed to be more receptive, adaptive and generative in order to meet the needs of stakeholders. Some of the characteristics of a healthy organization are (Johnson Rose and Katie Brunelle, 2004):

- Employee Morale: Relates to individuals valuing their positions in an organization; helps to keep track of what an organization is achieving and whether stakeholder needs are met
- Adaptability: Refers to the changing boundaries of an organization based on the rules and regulations which help an organization to be well structured
- Leadership: Good relations between employees and their management help the managers to keep track of employee performance and give constructive feedback to provide a scope for improvement; good leadership is an important characteristic of a healthy organization
- Teamwork: Teams working collaboratively to achieve a defined goal and collectively meet enterprise objectives
- Defined Structure: Helps to maintain integrity among the individuals involved in an organization
- Constant Learning/Emergence: This relates to giving importance to the feedback given by customers & to learn from experiences, which helps an organization to stay ahead. Adaptability and Emergence always help organizations to keep pace of technology and operational changes.

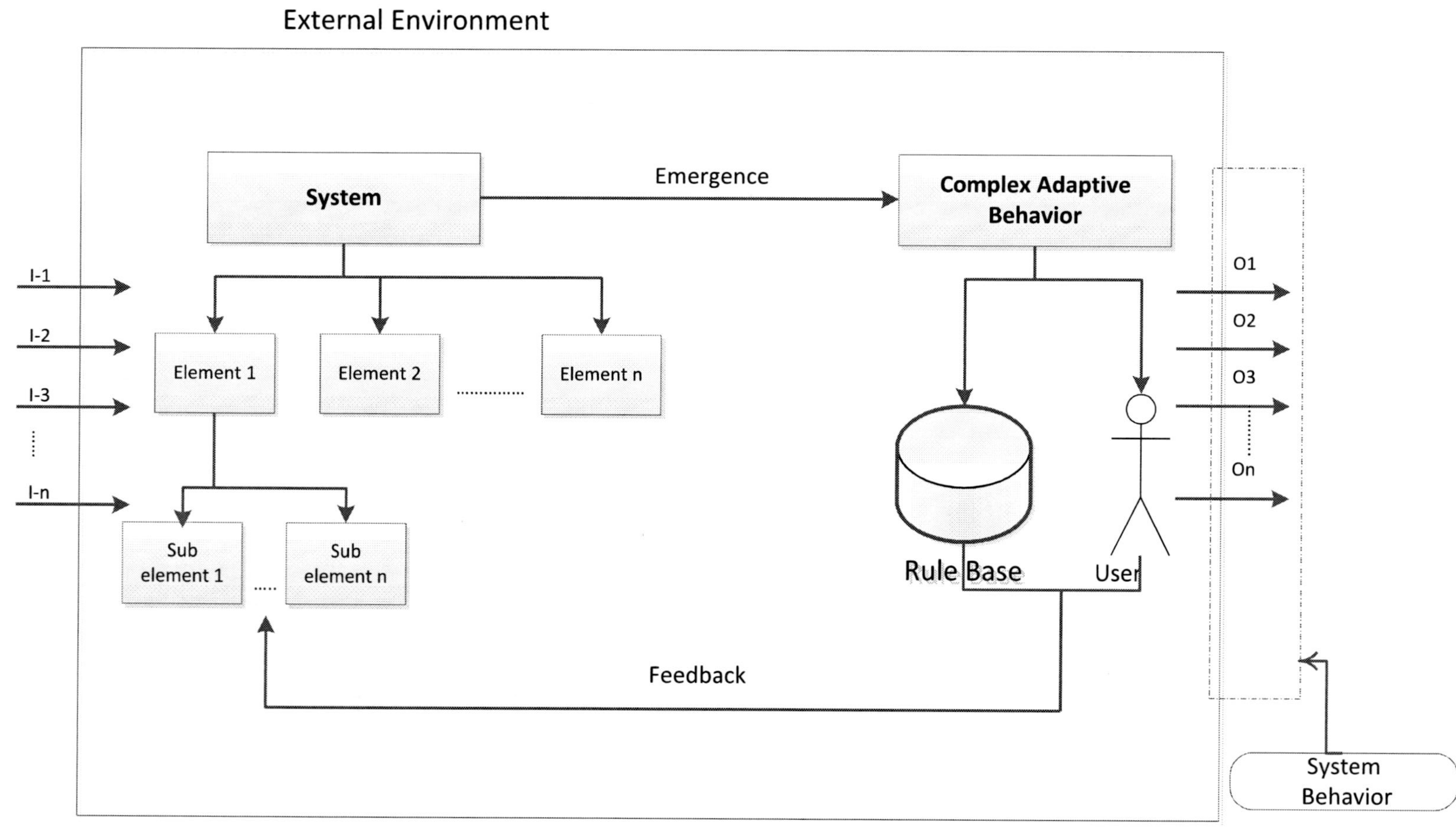

Figure 1. Representation of a complex adaptive system.

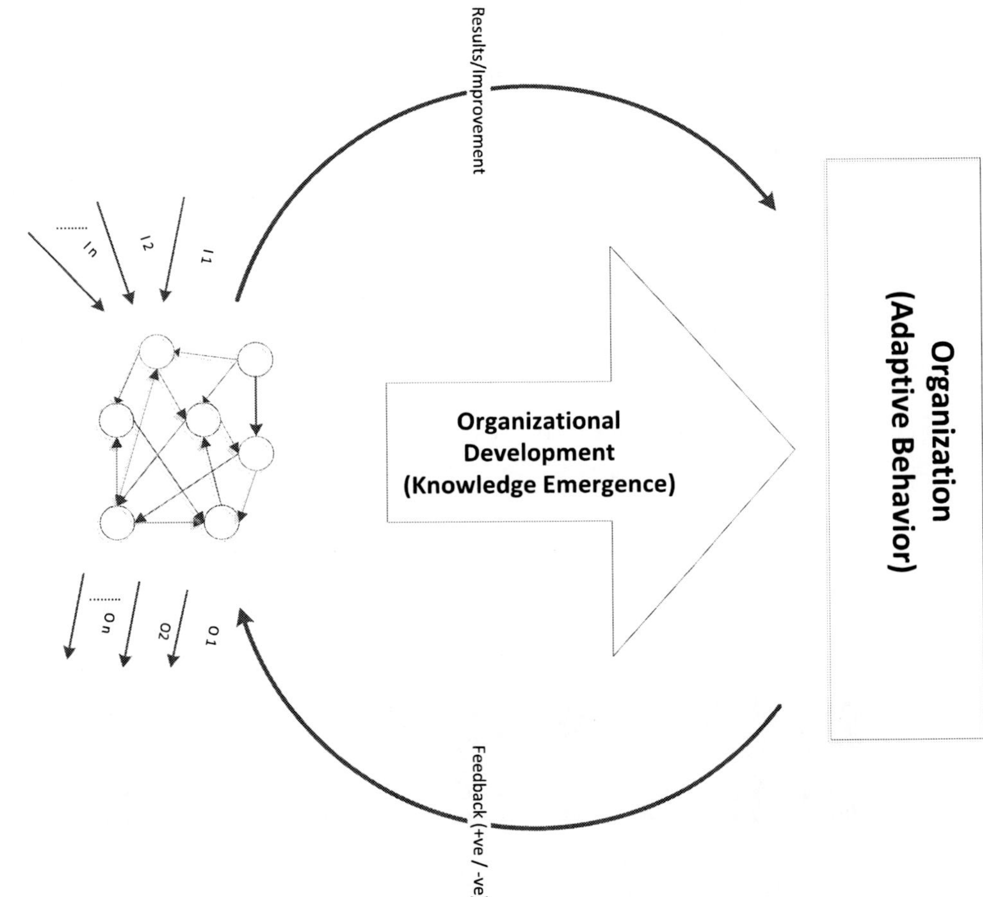

External Environment
(Customers, Products, Services, Metrics, Markets & Feedback)

Figure 2. Representation of organizations being knowledge emergent.

1.3. What Makes an Organization a Complex Adaptive System?

As illustrated in Figure 2, organizations have many inputs which make them self organizing and portray uncertain behavior. For knowledge emergent enterprises, understanding organizational structures and decision making processes along with the external influences on the enterprise is difficult. For an organization to be defined as complex, it should be able to demonstrate 3 types of equilibrium. According to Alligood et al., (1996), Stacey (1996), Stewart (1989), Thietart and Forgues (1995) the three equilibria are:

- Organizational Change and adaptability based on a positive feedback
- Maintaining Structural Symmetry after introducing a large organizational refinement along with introducing a shift in processes
- Structural Stability with many counteracting influences having a positive or a negative feedback periodically.

The Table1 developed compares the characteristics of an organization to those of a complex system and demonstrates what makes an organization behave as a CAS.

Figure 3. Types of fractals (Adapted from Wikipedia).

2. FRACTALS AND ORGANIZATIONS

A fractal can be defined as an object or an entity that exhibits the property of self similarity. Despite the fact that fractal objects come from the same elements, fractal object do not need to be exactly the same at all the scales. A few examples of fractal phenomena observed naturally are: Clouds, rivers, lightning bolts, snowflakes, blood vessels etc. The

Sierpinski triangle, Koch curve, and Cantor dust are few examples of complex and geometrical fractals.

According to Fryer and Ruis (2004), "Fractality" is termed as a way of thinking about collective behavior of distinct interacting units such as neurons and atoms with an ability to collectively evolve over a period of time. Their interactions over a period of time define a coherent behavior with emergent properties that can be observed only at a higher scale. Figure 4 illustrates an overview of an organizational structure scaling all the way from the top level to an individual.

Table 1. Dimensions of comparison between CASs and organizations

Complex adaptive systems	*Organizations*
1. Distributed control: *No single control mechanism. A defined structure is established for system functionality*	1. Defined structure: *A well defined organizational structure to maintain organizational integrity.*
2. Connectivity: *Interrelations between system elements and sub elements. These interactions of elements collectively define a system's behavior.*	2. Teamwork & leadership: *Relations with managers to keep track of performance and working collaboratively to achieve a common goal.*
3. Co-evolution: *Defines the change in behavioral patterns over a period of time.*	3. Constant learning: *Working towards a change in behavioral patterns over a period of time based on learning experience.*
4. Sensitivity & dependence: *Unpredictable behavior based on disturbances in operational environment.*	4. Adaptability: *Changing organizational boundaries based on rules and regulations.*
5. Emergent order: Being knowledge emergent and capable of self organization	5. Emergence: *Adapting to the pace of technological improvements to stay ahead based on feedback provided by customers.*
6. Individual element behavior: *Constituent system elements contributing to an overall system behavior based on their individual capabilities.*	6. Employee morale: *Individuals valuing their positions in organizations and working towards the enterprise vision.*

Organizational structures based on Figure 4, can be seen as complex fractals with nonlinear behavior. The interaction among a functional organization contributes to developing an ability to adapt with a changing environment. Such an organization can be characterized by the potential for self organization in a non-equilibrium environment. This potential helps organizations to evolve by random mutations and by transforming their internal structural models and processes, helping them to establish metrics, along with being knowledge emergent. Comparing an organization to a CAS, it can be seen that the individual system elements along the organizational hierarchy interact based on certain rules. These system elements are diverse both in capabilities and structure, helping them to adapt by being flexible to any changes, and hence, capable of gaining experience. Organizations, similar to fractal systems, evolve based on their past experiences and their past effects help to determine future trajectories. A great deal of uncertainty can be observed based on unanticipated environmental disturbances on their respective emergent structures.

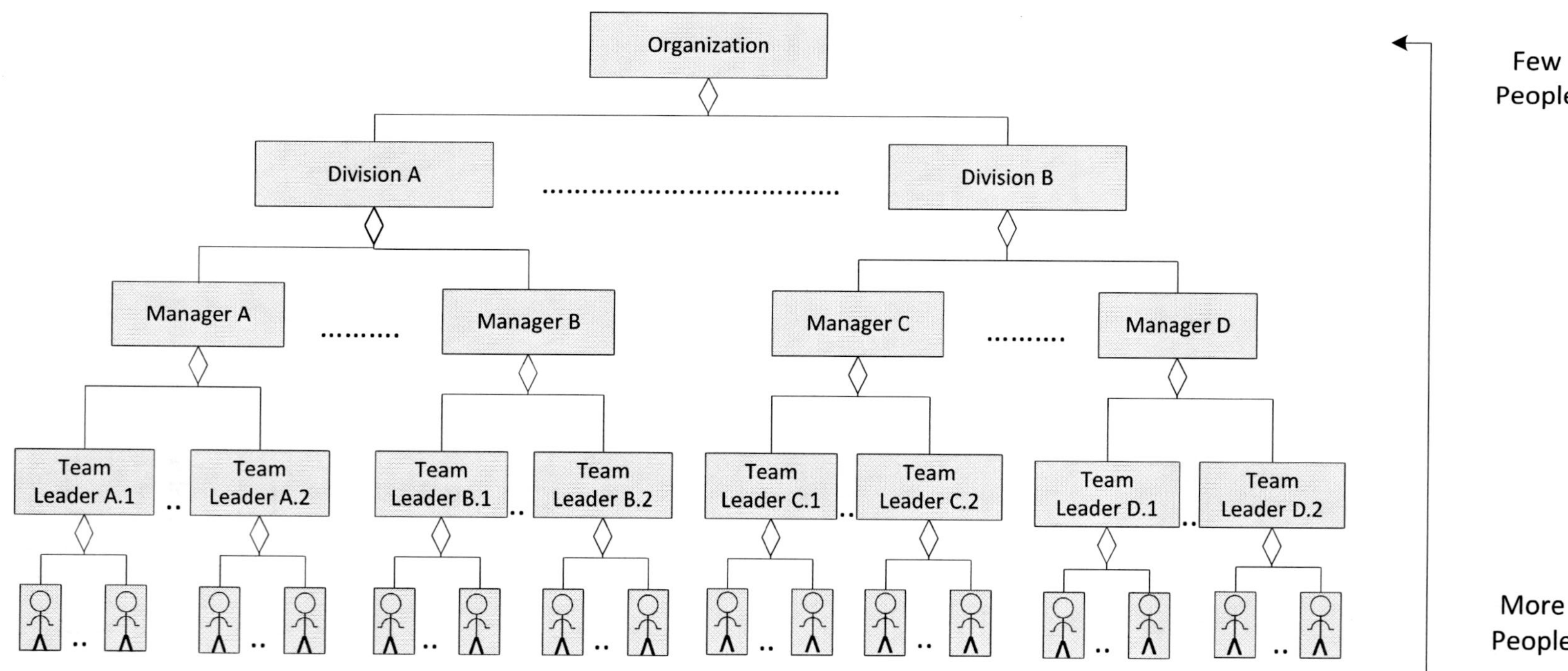

Figure 4. Organizational hierarchy as a fractal.

Similarities between fractals and organizations:

- Complex: Complexity in fractals can be defined by generating them in complex planes with an approximate scale of symmetry. In such fractals we observe many nested self similar structures. Examples are Mandelbrot and Julia sets. Also, organizations are considered to be complex structures with respect to Table 1.
- Collaborative: It defines the Interrelations between entities, system elements and sub elements. These interrelations and interactions among entities collectively define a structure. For example, the fractal patterns in cities can be observed to be tightly collaborative, giving rise to patterns in which a city is developed. Also, relations and interactions between employees and managers in an organization help them to work collaboratively to achieve a common goal.
- Structured: This implies a system model that handles governance and maintains balance between the external and internal system elements. This well structured governance helps to identify AS-IS and to determine the TO-BE state of the enterprise.
- Flexible & Self organized: Implies that there is no fixed structure or rule base. We observe constant reorganization of structural chains in organizations to be best fit in a considered environment.
- Nested Systems: Implies that the systems or entities are nested within each other resulting in many systems/entities as systems of smaller systems. For example, if we consider fractal patterns of cities, we find that they are nested systems. Similarly, a branch of an organization itself is a system with its customers, employee and suppliers. The same can be said of an organizational system as a whole working towards an established enterprise vision.

Thus, we can say that any organization as a whole can be seen as a collection of constituent self similar structures such as its divisions, departments, projects teams and so on. In each of these constituent systems, many employees either individually or in groups play important roles for organizational functionality and sustainability. Hence, the roles played by employees can be seen as integral toward defining an organization, and contributing to its behavior, success and constant emergence. Organizational changes can be observed as a result of a fractal shifts in the respective organizations with individuals being considered as individual fractals or as system elements.

3. DECISION MAKING BY INDIVIDUALS IN ORGANIZATIONS

Decision making can be defined as a logical process used to make an effective judgment from among available alternatives, and is an essential activity for an organization to survive.

According to Sven Ove Hansson (1994), Decision Theory is a systematic analysis of decision making. Mathematical and statistical methods are used by decision makers to pick a best alternative among those available. Decision making occurs frequently at various levels of an organization such as individuals, teams and management. The basic model of decision making along with rational assumptions to be considered is shown in Figure 6.

     Eric D. Smith and Aditya Akundi

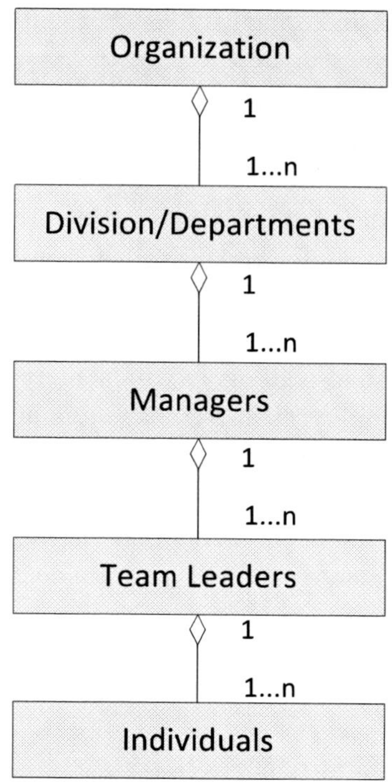

Figure 5. Higher level fractal organizational representation.

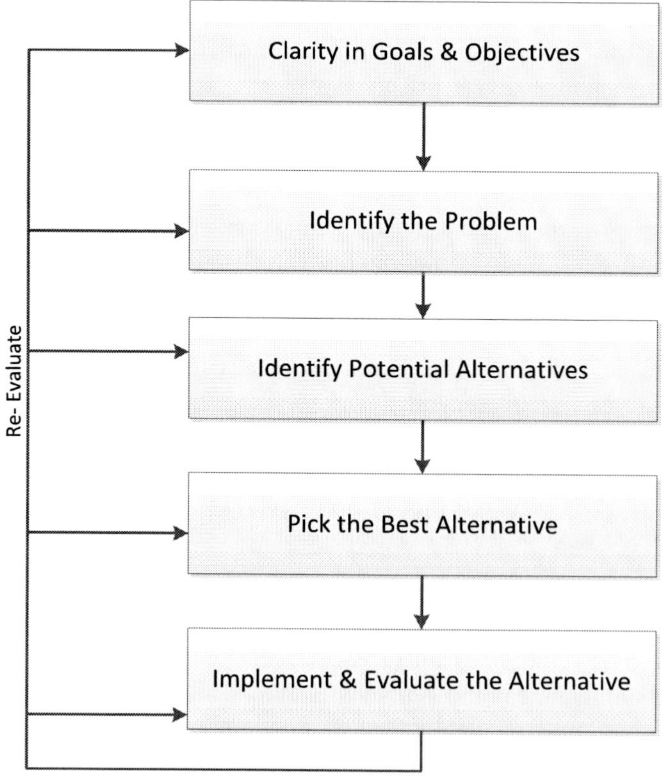

Figure 6. Rational decision making process.

There are many decision makers involved in all the associated levels of an organization. Clarity in goals and objectives serves as a guide for their decision making by establishing a criterion for identifying and evaluating alternatives. In short, with no clear established goals and objectives, there is no effective decision making.

Individuals set goals and objectives based on their own set of value system. Groups and organizations rely on group discussions, brain storming or other processes to establish goals and objectives. This results in different objectives established by the individuals and groups in an organization. Constant shift in the organizational objectives and goals is observed as a result.

Clear identification of the problem is a crucial part of any decision making process. Overlooking seriousness of a problem, focusing on pre-implied symptoms, and defining a solution to a problem without detailed analysis, are a few factors which contribute to incorrect identification of a problem. Cherrington (1989) states that, in this stage, decision makers usually tend to focus on a predefined course of action rather than exploring the real problem. Perceptual defense i.e. ignoring the information that may affect self morale is sometimes also observed.

Identifying potential alternatives includes having a clear understanding of the objectives and the problem statement. Developing best alternatives begins by using previously developed models deemed to be appropriate as a rule of thumb. A rational decision making process assumes that all the potential alternatives are identified to zero in on a best one. Sometimes in organizations if a satisfactory solution is found, potential alternatives are overlooked.

After identifying the alternatives, a set of criterion is established to compare the alternatives and to pick a best among them which is inclined to satisfy the problem identified and which satisfies the goals and objectives. Mathematical formulae and statistical analysis which identify the relationships are sometimes employed in this process. Although these methods are available, decision makers operate under the conditions of uncertainty, they strongly follow intuition to make such decisions.

Once the best alternative is selected, implementation plays an important role in organizations, as it addresses the behavioral changes required among the individuals to successfully implement the best alternative. Evaluation is also deemed necessary to keep track of the decisions made and to check if the results are consistent with the enterprise vision.

<table>
<tr><td><u>Rational Assumptions for a Decision Making Process</u>

• unambiguity in problem Statement
• know options/well defined alternatives
• Clear Criteria/preferences
• No Constraints (Cost, Time,..)
• Maximum yield</td><td style="text-align:center">Decision Making Process</td><td><u>How are Decisions actually made in Organizations ?</u>

• Bounded rationality
• Intuitive Decision Making
• Behavior decision theory
• Satisficing alternatives/ Incremental decision making
• Programmed Decision Making</td></tr>
</table>

Figure 7. Dimensions observed in decision making.

According to Cherrington, in reality, decision making by individuals is less orderly and systematic compared to the model explained. Figure 7 illustrates the different dimensions observed in decision making. In the following sections we first explore different possible cognitive biases and attributions observed that have influence in decision making and also examine how a typical individual's decision making processes is influenced by their intuition, rationality, possible biases and satisficing behavior.

4. COGNITIVE BIASES

The term cognitive bias was introduced by Kahneman and Tversky in 1972. Cognitive bias can be defined as a deviation from objective rationality observed in a decision, evaluation or a judgment as a result of human behavior, based on an independent adaptive set of tasks depending on a given context. In other words, cognitive bias is better defined as a phenomenon where in which, a genuine limitation in a thought process is observed depending upon situational contexts, false senses, and attributions.

- *How do humans actually reason?*
  - Using biases and heuristics: rules of thumb or a set of rules we use to simplify a cognitive task.
- *How should we actually reason?*
  - Using deductive logic, probability, statistics, decision theory and game theory.

We are often unaware of the heuristics we are using and the heuristics used are not designed to give optimal solutions but produce solutions that are "just good enough."

Bias is a predictable effect of using a given heuristic; this often produces a gap between how we ought to reason and how we in-fact reason. Example based on a study by Kahneman and Tversky (1974): when two groups of individuals were asked to determine what percentage of African nations are the members of the United Nations, they observed that for the group of subjects where the question was phrased "Is it higher or Lower than 10%?" the collective response was an average of 25%, and for the group of individuals where the question was framed "Is it higher or lower than 65%?" the response was on average 45%. We see that the second group estimated higher than the first group. Based on this study, it was observed that the response of these two groups was biased on an Anchorage effect, i.e. their estimates were anchored to the given percentage, where if they were give a higher number the response was higher, and when given a lower number their estimates were lower. So the idea behind the anchoring heuristic is that when people are asked to estimate a probability or an uncertain number, rather than using a complex calculation, they tend to use a reference point or an anchor to take a decision.

As critical thinkers there is a need to be aware of such processes that influence decisions, especially if decisions are systematically biased and are prone to errors, contributing to bad decisions. In terms of individuals involved in organizations, there is a need to look out for conscious manipulation and exploitation of these biases by the people involved in complex projects, or for influences in business where people need to act according to the organization's interests rather than their own.

4.1. Social Cognition and Attributions

Social cognition is a study by social psychologists who study how people interact and think in groups, and how groups influence individuals. Social cognition can be defined as a study of ways by which people process, store and remember information about others.

Attribution: a process through which an observer infers the causes of another's behavior. There are 2 types of attributions: *Dispositional Attribution* (when an observer attributes another person's behavior to internal states, i.e. Personality, thinking, behavior), and *Situational Attribution* (when an observer attributes another person's behavior to the environment, i.e. based on context, situation, environment, social structure). According to Jones and Harris (1967), to understand the importance of attribution, in a study participants were given an 'anti- Castro' or a 'pro-Castro' essay individually and were asked to give their insight on the respective writer's attitude. They observed that people tend to make correspondent attributions about their views on the writer even when they were priorly informed about the random assignment of the anti/pro essays to the writers; in-fact, the essays were written by the authors themselves.

4.1.1. Biases Observed in Attribution

Fundamental Attribution Errors

Fundamental attribution errors reveal a tendency to overestimate dispositional factors and underestimate situational factors in attributing the behavior of others. In other words, it is defined as a phenomenon when an observer tends to attribute others' behavior to their personalities but not to the situation/context of the behavior. Ex: the study by Jones and Harris (1967) as explained above where we see that the situational factor is completely ignored by the subjects.

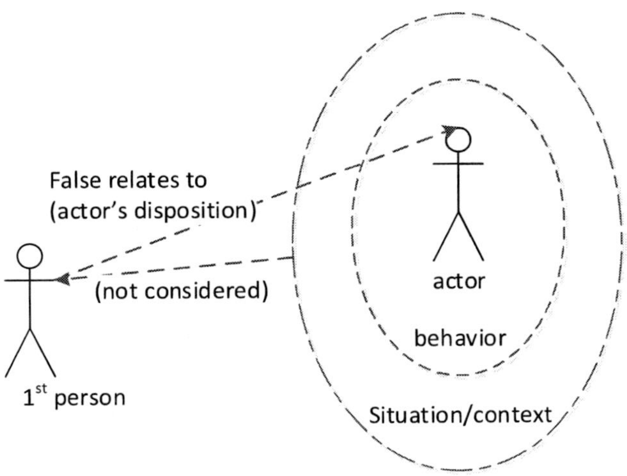

Figure 8. Fundamental Attribution Error.

Correspondence Bias

The Correspondence Bias is related to the fundamental attribution error; and is basically a tendency to contrive dispositional attributions. The Correspondence Bias can be defined as a tendency by an observer to assume a correspondence between an individual's behavior and underlying attitudes.

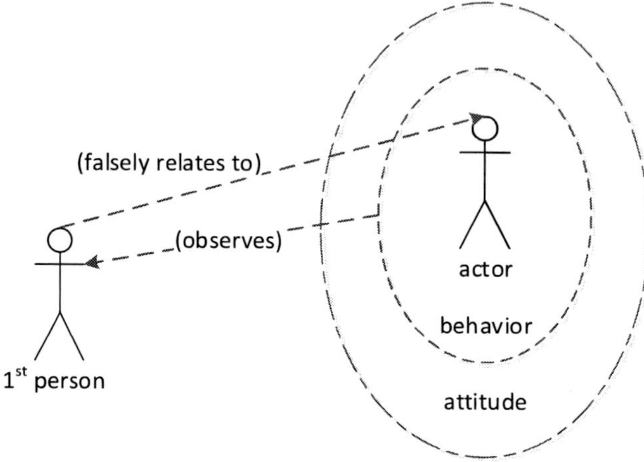

Figure 9. Correspondence Bias.

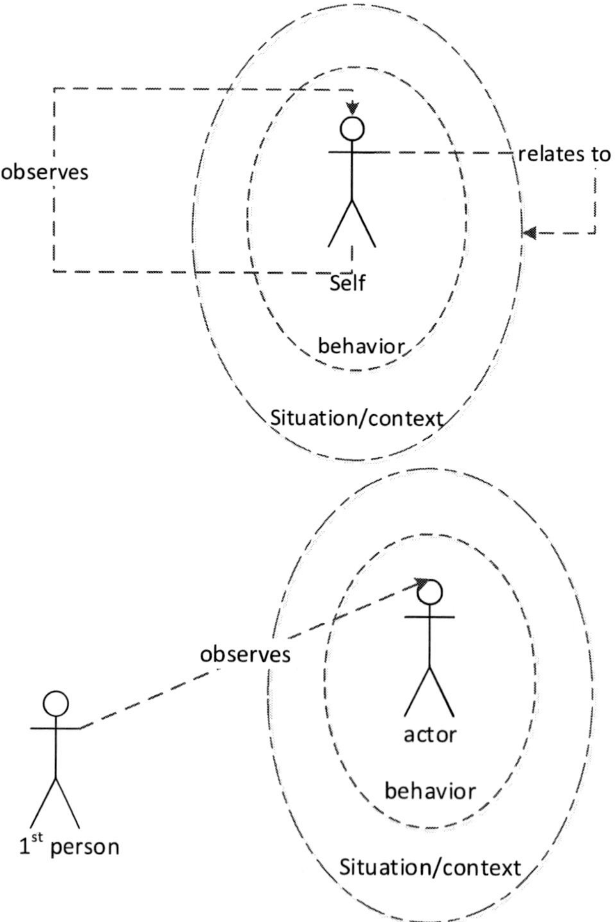

Figure 10. Actor-Observer Bias.

Actor-Observer Bias

Actor-observer biases reveal the tendency of an observer to see his/her behavior as an outcome of the context or the situation and others' behavior as a product of their internal states. This mechanism can also be known as Focalism, which is defined as a tendency of an observer to view others focally and to consider situational contexts when assessing one's own behavior.

To illustrate a subtle difference between different types of attribution biases explained, consider the following:

- Observer sees someone do something → DISPOSITIONAL ATTRIBUTION
- Observer does something good → DISPOSITIONAL ATTRIBUTION
- Observes does something bad → SITUATIONAL ATTRIBUTION

So How Should You Attribute?

According to Walter Mischel (1968), based on a study of effect sizes related to human personality, he observed that personality doesn't seem to be dependent on behavior. The correlation was observed to be around 0.40, i.e. behavior explains only 16% of the variance in personality.

As people have domain-specific personality types (situations, person-in-time interactions, circumstances, etc.), based on studies by sociologists and personality researchers, it is advised to consider different situations to equalize personality types before association of attributes.

Bounded Rationality

According to Simon (1999) and Jones (1999), the notion of bounded rationality has its origins in political science. It is defined as a phenomenon where individuals are limited by the information available, limiting their thought process to a finite amount of time available for decision making. In other words, when decision makers are limited as to resources for finding an optimal solution, they use bounded rationality to simplify the complexity of the task of decision making.

Intuitive Decision Making

Intuition is defined as a process of formulating ideas or inputs without knowing their roots. A few examples of where intuition plays an important role are: selecting a right car to buy, evaluation of a job offer, decision making in educational pursuits, selecting a meal before eating, and so on. According to Albert Einstein 'There is no logical way to the discovery of these elemental laws. There is only the way of intuition, which is helped by a feeling for the order lying behind the appearance'. But if decision makers use intuition in organizational decision making processes, there may be a significant negative influence on the decisions being made which may not always align with the overall enterprise vision.

Incremental Decision Making

Decision makers knowingly avoid making drastic changes to policies or procedures in order to successfully implement decisions. These decisions are mitigated by sometimes compromising with the stakeholders, and they tend to focus on short-term goals rather than long-term.

According to Schwenk (1982) theorists in the field of strategic management and decision making have pointed out that human cognitive limitations might have an effect on decision making processes, based on the notion of bounded rationality. Schwenk cites a statement from Nisbett and Ross (1980) adapted from Kanouse (1972) which explains:

"Individuals may be primarily motivated to seek a single sufficient explanation for any event, rather than the one that is the best of all possible explanations. In other words, individuals may exert more cognitive effort in seeking an adequate explanation when none has yet come to mind than they do in seeking for a possibly better solution when an adequate one is already available. This bias may reflect a tendency to think of unitary events and decisions having unitary causes; individuals may assume, in effect, that no more than one sufficient explanation is likely to exist for a single phenomenon" (Kanouse, 1972, p.131).

Table 2 below illustrates and helps to understand how different types of biases influence the different stages of decision making processes as based on Figure 6, thereby affecting the quality of decisions made.

Table 2. Possible scenarios where bias can be observed in decision making

Stages in decision making process	Possible bias that could be observed	Influence and the effect of bias in decision making
Goals, Objectives and Problem identification	Prior hypothesis	Ignoring evidence based on possible prior knowledge, probable gaps not properly analyzed, thus introducing uncertainty.
	Prior success	Arriving to conclusions based on prior, similar, and successful projects (either industrial or in academia).
Identification of Potential alternatives	Personal intuition	Precocious exclusion of alternatives
	Anchoring effect	Developing alternatives based on self-made reference points.
	Incremental decision making	Developing alternatives for satisfying short term goals.
	Bounded rationality	Not considering complexity. (In other words, developing alternatives by simplifying complexity based upon time and resource constraints)
Alternative selection, Implementation and Evaluation	Fundamental attribution error	Taking decisions by inferring to or based on customer (stakeholder) behavior
	Personal intuition	Taking decisions based on personal judgment by not considering the possible risk factors
	Certainty effect	Reducing evaluation criteria based on anchored beliefs.

We now introduce a few models helpful for bias evaluation such as: Force Field Analysis, Balance theory for Cognitive Consistency, Spiral Model for Bias Identification, and the V-Model for Decision Planning. These change agent models can help individuals to understand possible biases as explained in Table 2, and can help them to re-evaluate decisions being made before finalizing them.

5. FORCE FIELD ANALYSIS TO UNDERSTAND THE BEHAVIOR OF DECISION MAKERS

Force Field Analysis is a significant contribution in the fields of organizational change, social science, and social psychology as developed by Kurt Lewin. It provides a framework for understanding the factors that influence a situation, which are either geared towards a specific goal or away from it. This technique helps to identify and understand the attributes which are against a behavioral change to address the problem. Figure 11, developed based on the principles of force field analysis, illustrates the forces "for" and "against" effective decision making by decision makers in organizations.

The Attributes identified in any complex system such as an organization, allow perturbations, i.e. the forces against effective decision making, to percolate across a decision network significantly. In order to understand the impact of such attributes, we suggest a mathematical model adapted from Aufderheide, Rudolf, Gross and Lafferty (2013), as shown below.

5.1. Impact Evaluation of the Forces against Effective Decision Making

Every organization follows a pre-determined set of rules which govern a complex decision making processes based on the situation or a project accordingly. When these decisions are being made by the individuals specifically assigned to the projects, there is a possibility of cognitive bias, personal intuition or professional experience having a significant influence.

Consider a standard decision making process used in organizations and let them be characterized by the set of variables $x_1, x_2, x_3, \ldots\ldots\ldots, x_n$. The perturbations, i.e. the cognitive biases, personal intuitions and other forces against effective decision making which influence individuals in a decision making process, can be characterized by the variables $y_1, y_2, y_3, \ldots\ldots, y_m$. Based on Russell (2003), and March & Simon (1993), decision makers tend to reduce the problems based on their own assumptions and mental models, along with biases. Also, it is observed that the solutions they zero in on are interpreted as a way to support their experience and personal beliefs. Hence a decision making process over a period of time can be represented with the help of a differential equation such as

$$\tfrac{d}{dt}X_i = F_i\ (x_1, x_2, \ldots, x_n, y_1, y_2, \ldots\ldots, y_m) \tag{1}$$

This is the case where there is no effect of perturbations on the decision making process, i.e. a state of equilibrium be represented as X_i^* and Y^*, representing the set of variables characterizing the external variables. Considering the effect of perturbations on the decision making process, the equilibrium state X_i^* can be expressed as a function of Y^*

$$X_i^* = X_i^*\ (Y^*) \tag{2}$$

We can define the impact I of a variable Y^* on X_i^* as the change of X_i^* per Y^*

$$I = \frac{\partial X_i^*}{\partial Y^*} \tag{3}$$

In other words, the impact can be calculated by solving equation (1) for stationary solution $X_i^*(Y^*)$ and computing Equation (3). From this, though we get a general impact of perturbations on the decision making variables, if implicitly found, it would be helpful to understand the effect of each individual perturbation on the process. To address this, we formulate an evaluation based on the implicit function theorem.

According to this theorem, a stationary implicit function:

$$F\ (y, x_1, x_2, x_3, \ldots, x_n) = 0$$

Can also be written as:

$$F\ (y, x_1, x_2, x_3, \ldots, x_n) = Z \ assuming \ Z = 0$$

Taking derivatives on both sides of the equation:

$$dZ = F_y dy + F_{x_1} dx_1 + F_{x_2} dx_2 + \cdots \ldots \ldots \ldots + F_{x_n} dx_n$$

For $Z = 0\ we\ have\ dZ = 0$, this implies that the equation above can be written as:

$$0 = F_y dy + F_{x_1} dx_1 + F_{x_2} dx_2 + \cdots \ldots \ldots \ldots + F_{x_n} dx_n$$

To find the change in y with respect to x_1 substitute $x_2, x_3, \ldots \ldots, x_n = 0$
Thus we have:

$$F_y dy = -F_{x_1} dx_1$$

i.e.

$$\frac{dy}{dx} = -\frac{F_{x_1}}{F_y}$$

Using partial derivatives to find the rate of change, the equation above can be written as

$$\frac{dy}{dx} = -\frac{F_{x_1}}{F_y} = -\frac{\frac{\partial F}{\partial x_1}}{\frac{\partial F}{\partial y}}$$

Based on the above formulation for X_i, being an implicit function, assume a stationary solution where a decision taken in a life cycle stage of a project in any organization remains the same over a period of time represented by:

$$F_i\left(X_i^*, Y^*\right) = 0 \tag{4}$$

According to the implicit function theorem, we can write the impact matrix:

$$I = \frac{\partial X_i^*}{\partial Y^*} = \frac{\frac{-\partial F_i}{\partial Y^*}}{\frac{\partial F_i}{\partial X_i^*}} \tag{5}$$

It can also be expressed as

$$I = -KJ^{-1}$$

where:

$$K = \frac{\partial F_i}{\partial Y^*} = \frac{\partial}{\partial Y^*}\left(\frac{dX_i}{dt}\right)$$

and

$$J = \frac{\partial F_i}{\partial X_i^*} = \frac{\partial}{\partial X_i^*}\left(\frac{dX_i}{dt}\right)$$

Here, the matrix J defines the derivatives of F_i with respect to the established parameters which gives the internal effects of the variables and the matrix K captures the direct impact of the perturbations on the variables characterizing the decision making process.

6. USING BALANCE THEORY TO MAINTAIN COGNITIVE CONSISTENCY

Balance theory, as proposed by Fritz Heider, is a motivational theory which helps a decision maker to keep track of an attitude change. This theory is based on cognitive consistency, which is a motive to achieve psychological balance over a period of time.

In organizations, it is always difficult to understand the behavior of management oversight and individual decision makers if they are not in inclination with the established enterprise vision. Eventually the customer satisfaction of an organization may decline, if the decision makers and the management are inconsistent to the enterprise vision with their decisions. Figure 12, geared towards organizational decision making, is based on Heider's P-O-X model.

In relation to the decision makers, management oversight and enterprise vision, the model is said to be in balance when the multiplicative resultant of the signs on the triangle is a positive solution. Considering the decision makers to be the focal point in this model, the relation between the decision makers, management and the enterprise vision is determined based on their respective attitude towards each other. If the multiplicative resultant of the signs of these relationships towards each other is positive, a balanced state is said to be achieved. There are 8 possible combinations in this model which would help to determine the logical and relational characteristics.

Assuming the enterprise vision to be in favor of positive organizational development, it would be helpful to regulate and keep track of decisions being made in an organization. Checking for their logical balance towards the enterprise vision and providing a constructive feedback, helps to motivate the decision makers involved towards a positive behavioral change.

7. SPIRAL MODEL FOR INTEGRATED BIAS IDENTIFICATION

This model, as shown in Figure 13, illustrates a development and iterative process combining five different methodologies intended to identify and analyze possible biases by decision makers, with a decision taken being the focal point. This spiral model is intended for individual decision makers, and is geared towards helping them to identify the possible bias that would have influenced decisions taken in complex projects. This model is developed based on combining the idea of an iterative incremental approach each time a loop is completed successfully.

Starting with a *Decision being* made, a loop around the spiral goes through: *Checking for bias, Impact Evaluation, Gathering Evidence,* and *Providing Constructive Feedback* along with *ensuring the change* required. This model includes the identification of possible biases once a decision is made. The impact on the decision making variable with respect to the identified perturbations would then be calculated. Based on the impact identified, the decision variables with high impact factors are to be reevaluated based on the evidence gathered. Providing constructive feedback is seen as an important learning process to help the decision makers in order to understand the influence of their bias on the decisions made. Documenting this process throughout this cycle helps to identify the types of bias frequently being observed and develop strategic action plans to avoid these.

8. V-MODEL FOR STRATEGIC DECISION PLANNING

The V-model illustrated in Figure 14 represents a development process for a strategic plan, for effective organizational functionality adopted from the roots of the Systems Engineering V-Model. This model demonstrates the relationships between each phase of the cycle associated with effective decision making and re-evaluation. The horizontal axes are considered to be the most important aspects of the model, representing the decisions being made in organizational projects.

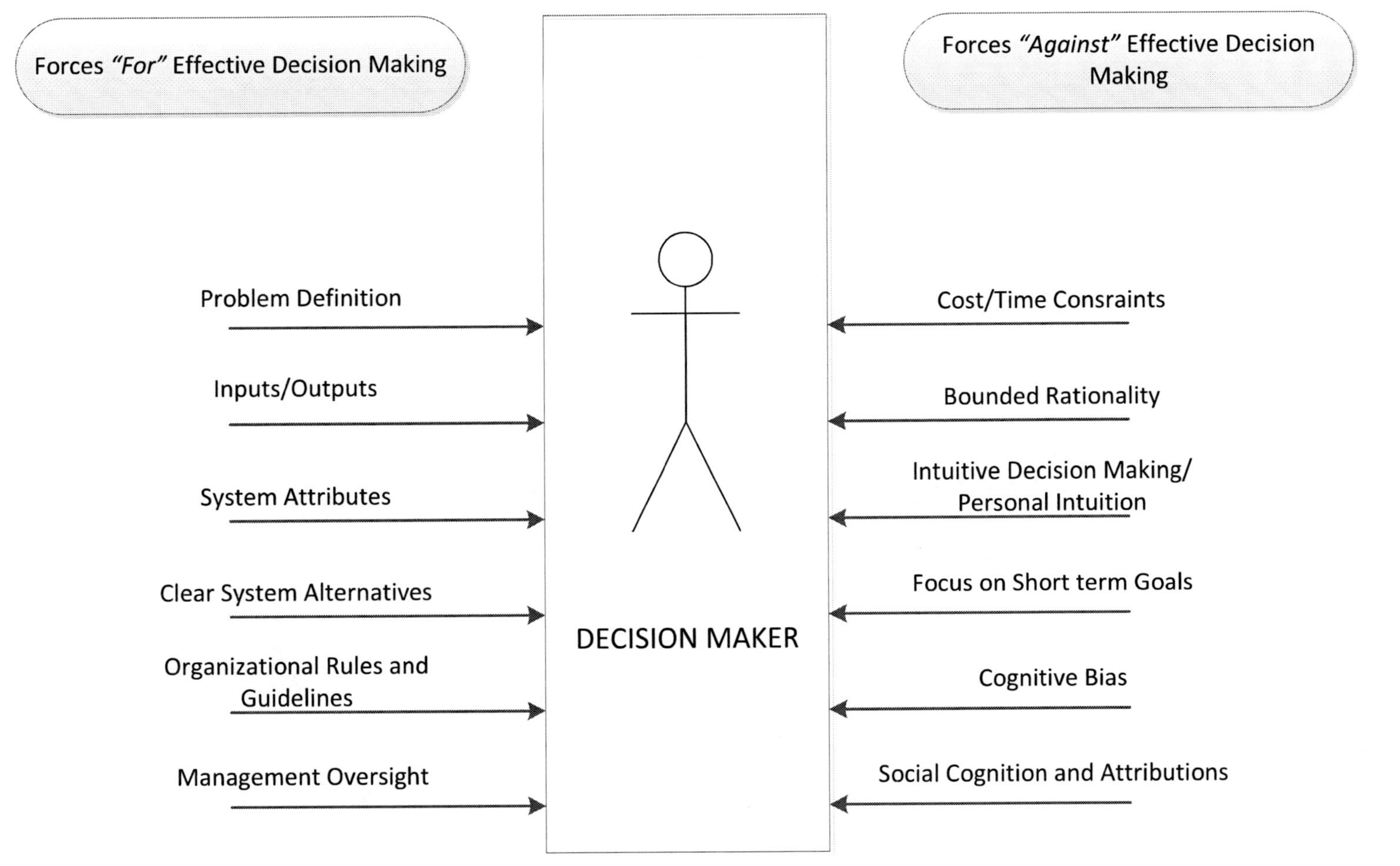

Figure 11. Decision making attributes identification using force field analysis.

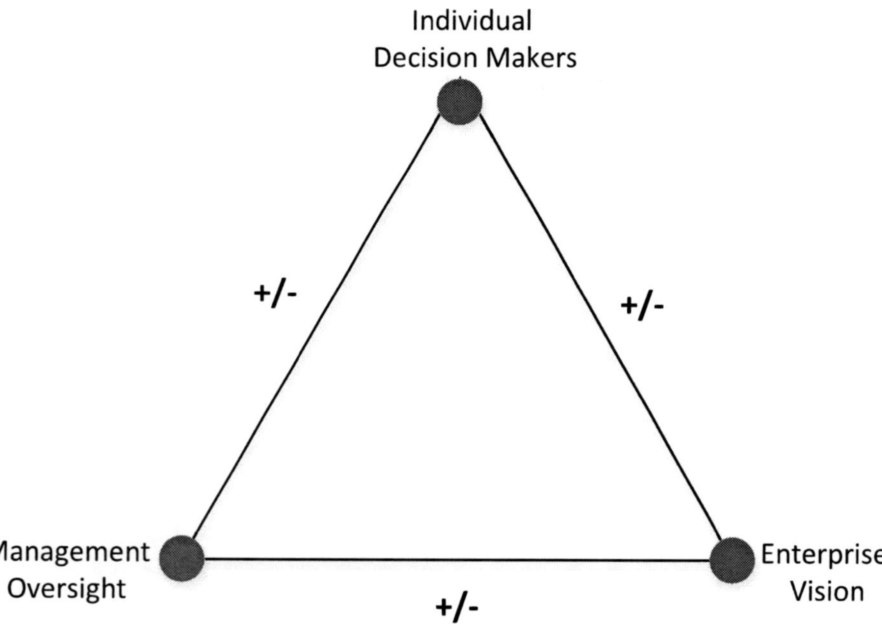

Example:

A Balance is maintained if:

- If Management takes the decisions inclined towards the Enterprise Vision.
- If the Individual decision makers anchor their decisions based on the management oversight.
- If the decision makers seek balance towards Enterprise Vision.

Figure 12. Balance model to maintain cognitive consistency.

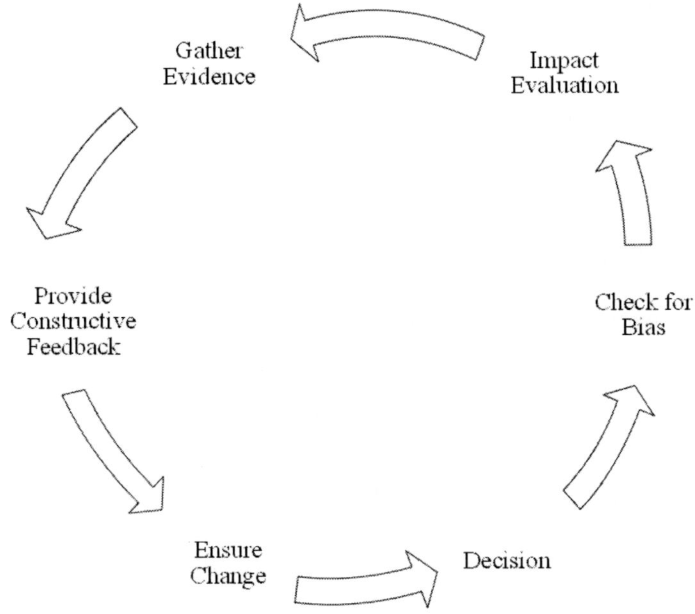

Figure 13. Spiral model for bias identification.

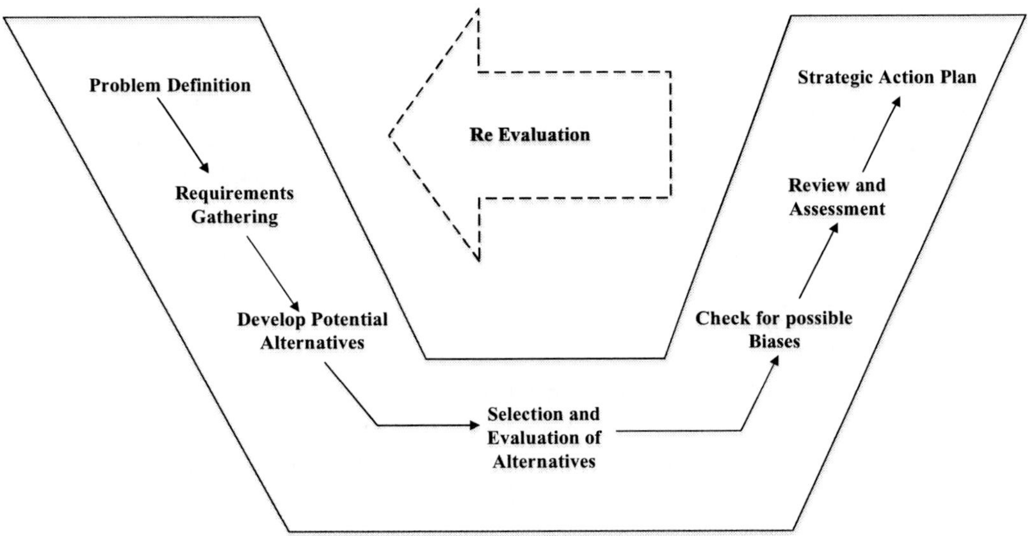

Figure 14. V-model for strategic decision planning.

Problem Definition

This phase signifies establishing the groundwork required to understand the problem statement. It includes identification of system needs prior to system analysis for decision making

Requirements Gathering

This phase, also known as requirements analysis, encompasses the tasks required for determining the constraints, needs and functionality of a system to address a specific need or problem statement.

Developing Potential Alternatives

In this phase, identifying potential alternatives shall include a clear understanding of the problem statement required to develop best alternatives, either based on previous models or on the established attributes in the previous stages.

Selection and Evaluation of Alternatives

This phase establishes guidelines to identify a best alternative based on important system requirements that helps to differentiate the alternatives, along with mathematical and statistical analysis.

Checking for Possible Bias

This phase is geared towards top level management and the team leaders involved in effectively identifying the possible biases in the preceding phases. This is because, though organizational guidelines are available, decision makers under the conditions of uncertainty may strongly follow intuition to make decisions.

Review and Assessment

Review and assessment of the prior phase helps to identify the critical aspects of decision making where biases play a role, if any. Also, the impact factor of each possible bias is to be calculated so that the most influenced decision variable can be addressed first.

Strategic and Action Plan

A strategic plan should then be laid out by the management by providing guidelines on how to avoid biases while decision making. This plan is to be ensured by developing a learning process among the critical decision makers, for them to understand and act accordingly based on their individual constructive feedbacks provided.

9. BOTTOM-UP PYRAMID APPROACH TO INTRODUCE BEHAVIORAL CHANGE

A behavioral change in an organization reflects on many significant effects, and on how a problem or a project is perceived by decision makers. A behavioral change sometimes may be apparent on organizational growth, when a gap is observed on how the decision makers make critical decisions in complex projects and how they are required to be made based on given organizational guidelines. Such a change may require a large scale gap analysis among the decision makers to understand how the decisions are being made and how they should ideally be made.

This problem can be addressed by applying a few change agent models such as: Balance theory for Cognitive Consistency, Spiral Model for Bias Identification and the V- Model for Decision Planning, as explained along the organizational structure.

These models can be used to motivate the decision makers to incline their thought processes to the overall enterprise vision, thereby contributing to organizational stability, change efforts, and effectiveness. A regular approach should be followed in organizations to introduce a behavioral change, where the top management identifies a behavioral change, the middle management often initiates a plan to implement the change, and supervisors or team leaders implement the plan by facilitating the resources needed be the employees. In this approach, where the big picture breaks down into smaller segments, if a segment in the hierarchical chain fails to elucidate the elementary change parameters, the essential change required may not be observed.

Contrary to a top down approach, we suggest a bottom up approach where the change agent models are introduced among the individual decision makers as shown in Figure 15. Any given change needs to be initially implemented among the individual decision makers, who collectively lead to more efficient organizational behavior. This change process, based on the spiral model, should ensure that the project team leads or the supervisors keep a track of the decisions being made along with calculating the impact of individual biases on the decisions. Providing constructive feedback based on the observed results would help to tune the thought process towards the enterprise vision while making critical decisions.

Once the desired behavior is identified, a balance theory approach would gear the team leads/supervisors and the decision makers to maintain a balance among the collective decisions taken towards a common enterprise vision. This approach helps to perceive the gaps people have that constrain them from the required behavioral change.

With the gaps identified, the top management can develop strategic plans to implement competency models for reinforcing a behavioral change among the individuals with the help of enterprise oversight. Change agent models similar to the V-model for strategic decision planning help to unfold the AS-IS organizational behavior in order to develop a strategic plan for identifying gaps to mentor and empower the decision makers toward a required change.

The model, if systematically addressed, would ensure the individuals to develop competency and avoid biases in decision making along a planned and controlled fashion, thereby increasing performance.

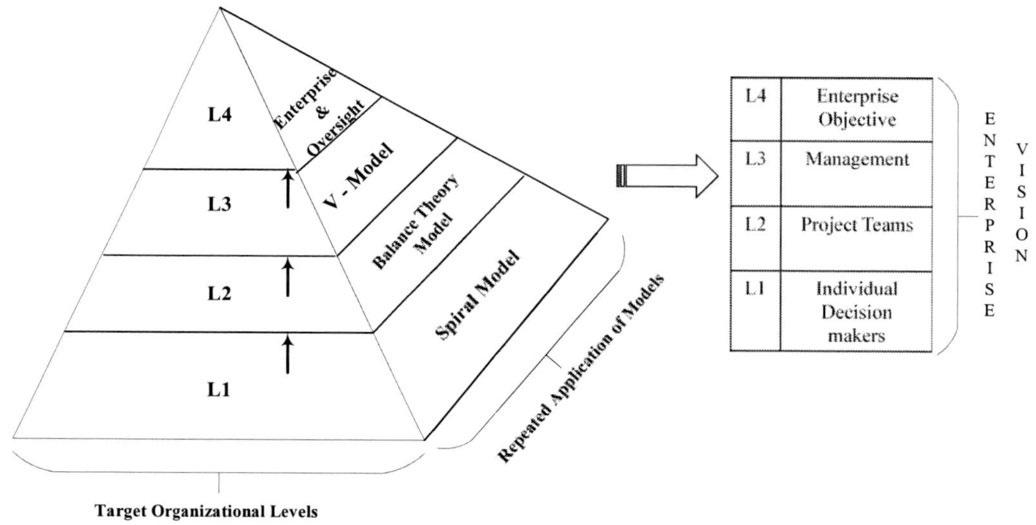

Figure 15. Pyramid approach for introducing behavioral change in organizations.

REFERENCES

Alligood, K., Sauer T. D., Yorke J. A. 1996. Chaos: *An Introduction to Dynamical Systems.* Springer-Verlag, New York.

Brunelle Katie, (2004), *Characteristics of a Healthy Organization*; University of Minnesota Duluth.

Berkowitz, W. R, and Wolff, T. J. (1999). *The spirit of coalition building.* Washington, D. C.: American Public Health Association.

Chan, Serena. *Complex Adaptive Systems* <URL: *Web.mit.edu.*> ESD.83 Research Seminar in Engineering Systems, 6 Nov. 2001. Web.

Cherrington, David J. *Decision Making in Organizations.* Organizational Behavior: The Management of Individual and Organizational Performance. Boston: Allyn & Bacon, 1989.

Fryer Peter and Ruis Jules (2004), *A brief description of 'Complex Adaptive and Emergent Systems'* (CAES), Edinhoven.

Holland, John H., (2006). Studying Complex Adaptive Systems, *Journal of Systems Science and Complexity* 19 (1): 1-8. < URL: >.http://hdl.handle.net/2027.42/41486>.

Johnson, Rose, and Demand Media. *The Top 10 Characteristics of a Healthy Organization* <URL: Smallbusiness.chron.com>.

Jones, E. E.; Harris, V. A. (1967). The attribution of attitudes. *Journal of Experimental Social Psychology* 3 (1): 1–24.

Jones, Bryan D. (1999). *Bounded Rationality. Princteon University.* Arjournals. annualreviews.org.

Kanouse, D. E. *Language, labeling, and attribution.* In E. E. Jones, et al. (Eds.). Attribution: Perceiving the causes of behavior. Morristown, New Jersey: General Learning Press, 1972.

Mischel, W. (1968). *Personality and assessment.* New York: Wiley.

Nisbett, R. & L. Ross, *Human inference.* Englewood Cliffs, New Jersey: Prentice-Hall, 1980.

Nagy Jenette, *Organizational Structure: An overview* <URL: http://ctb.ku.edu/ en/ tablecontents/sub_section_main_1092.aspx>.

Stacey, R. 1996. *Complexity and Creativity in Organizations.* Berrett- Koehler Publishers, San Francisco.

Stewart, I. 1989. *Does God Play Dice: The Mathematics of Chaos?* Basic Blackwell, Oxford, UK.

Simon H. A. 1999. *The potlatch between political science and economics.* In Competition and Cooperation: Conversations with Nobelists about Economics and Political Science, ed. J Alt, M Levi, E Ostrom. Cambridge, UK: Cambridge Univ. Press.

Schwenk, Charles R. *Cognitive Bias in Strategic Decision-making.* College of Commerce and Business Administration, Bureau of Economic and Business Research, University of Illinois at Urbana-Champaign. N.p., Apr. 1982. Web. http://openlibrary. org/books/ OL24623170M/Cognitive_bias_in_strategic_decision-making.

Thietart, R., B. Forgues. 1995. Chaos theory and Organization. *Organ.Sci.* 6(1) 19-31.

Tversky, A., and D. Kahneman. *Judgment under Uncertainty: Heuristics and Biases. Science* 185.4157 (1974): 1124-131. Print.

Weaver, Warren (1948). Science and Complexity. *American Scientist* 36: 536 (Retrieved on 2007–11–21.)

In: Advances in Systems Engineering Research
Editors: Elena Fermi and Adam Lamberti

ISBN: 978-1-62948-310-8
© 2013 Nova Science Publishers, Inc.

Chapter 5

APPLICATION AND VALIDATION OF SYSTEMS ENGINEERING METHODS AND TECHNIQUES IN PRACTICE

*Gerrit Muller**
Buskerud University College, Kongsberg, Norway

ABSTRACT

The Systems Engineering Body of Knowledge (SEBoK) is often described as a set of best practices. However, literature provides little substantiation for applicability and value of the captured methods and techniques. A research question is when these methods and techniques are applicable, and how much they improve performance of project teams. Research challenges are to collect data in the real world -where most influencing factors are not under control of the researchers- and to cope with the heterogeneity of the field of application. For example, the system-of-interest can be an integrated circuit or a space station; it can be well bounded or embedded in a complex socio-technical environment.

This chapter reports our research in an industrial cluster, where we have sampled the application of various systems engineering methods and techniques in a variety of domains. We discuss the research methodology, and we show a number of emerging trends after taking 67 samples in industrial practice in Kongsberg. Examples of applied methods and techniques are requirements management, interface management, modeling, concept selection, and A3 architecture overviews and reports.

The amount of samples is too low to answer the main research question. We observed that application of methods and techniques from the body of knowledge is far from trivial. Current practice and culture in the organization constrain application of methods and techniques. In many cases, there is a clear potential for improvement. The researchers selected methods or techniques for application based on observed problems in the past. The evaluation often shows interest and a certain enthusiasm from the stakeholders for these methods and techniques. However, this enthusiasm in itself is insufficient to trigger changes; more pulling force is required to introduce these techniques at a broader scale.

---

* Corresponding author: Email: Gerrit.muller@hibu.no.

INTRODUCTION

Researching Systems Engineering

The Systems Engineering Body of Knowledge (SEBoK)[1] is often described as a set of best practices. However, literature provides little substantiation for applicability and value of these practices. Best practices are often captured as methods and techniques. Our research question is when these methods and techniques are applicable, and how much they improve performance of project teams.

The Status of Systems Engineering as Discipline

Systems Engineering is a young discipline that is mostly practiced "in the field", e.g. in industrial companies. Dixit and Valerdi [2] state that many systems engineers do their work based on experience. An experienced group of INCOSE editors filled the Systems Engineering Body of Knowledge (SEBoK) [3] with a rich collection of best practices. Boehm, Valerdi, and Honour [4] research a hot question in the systems engineering community: what is the return on investment of applying systems engineering or what is the value of applying systems engineering. More general the academic question is "how do we know the effectiveness of these best practices?" More specific we like to be able to relate effectiveness to the context; what practice is appropriate in what circumstances?

Challenges of Researching Systems Engineering

Research challenges are to collect data in the real world -where most influencing factors are not under control of the researchers- and to cope with the heterogeneity of the field of application. For example, the system-of-interest can be an integrated circuit or a space station; it can be well bounded or embedded in a complex socio-technical environment. Additional challenge is that every experiment is unique; repeating an experiment with different parameters in the same setting is not possible.

In this chapter, we report the results of field research to study effectiveness of systems engineering methods and techniques in practice. Field research facilitates researchers to observe effectiveness when practiced in the field. Dominant research approaches to study in the field are action research [5] and industry-as-laboratory [6,7]. The researcher combines active participation in the systems engineering activities with the researcher role. The researcher is wearing two hats in these research approaches: as systems engineer and as researcher. The attitude of the systems engineer is result oriented and cooperative. The attitude of the researcher is questioning and challenging. Where the systems engineer will promote the use of a proposed method or technique, the researcher needs to question its validity.

Researchers in systems engineering are typically educated in the technical domain. However, effectiveness of systems engineering depends largely on human aspects, such as competence and behavior of individual stakeholders, social interaction between stakeholders, political circumstances, organization and governance, and many more. Research in systems engineering has to build on available scientific methods, both technical, as well as from the social sciences. Bhattacherjee wrote a good introduction in social science research [8].

Researchers can use a collection of case studies to evaluate theories or to form theories. However, the reverse approach, starting with case studies and eliciting theories is possible too. In the social sciences, Glaser and Strauss [9] laid the foundation of grounded theories. Strauss and Corbin have evolved the approach; see their overview in [10]. Researchers applying grounded theory collect data and extract key points called codes. Next, they cluster codes into concepts, form categories, and finally formulate a hypothesis. In other words, this approach is bottom-up, starting with observations and gradually crystallizing a theory. This way of working seems quite appropriate in a young field as systems engineering with many similar challenges as the social sciences. Charmaz [11] recommends memo writing as a technique in this process from data to concepts. We grow this collection of case studies to facilitate the formation of grounded theories by performing a kind of meta-analysis on them, for example as proposed by Glaser, Strauss, and Charmaz.

Research Methods in Industry as Laboratory

The nature of action research or industry as laboratory complicates this research. The way of doing this type of research requires research methods that fit in the socio-technical context. In a previous publication [12], the author describes the methods that researchers in our program have used in this research so far.

Figure 2 shows a context diagram of the research and many entities that are relevant for the research. If we study how effective a concept selection technique is, then we have to observe and describe the people using them, the process in which the concept selection is applied, the stakeholders and their concerns and objectives, the artifacts used, and the concepts themselves. Researchers have to strive for understanding the causes of success or failure. Is the technique, the embedding process, the people, or some other factor the cause?

Research methods assist in data collection and evaluation. These methods can range from free format and relatively informal to a standardized format with more formalized definitions. Examples of free format data collection are open interviews, sketches, and log observations. Examples of standardized methods are formal models, structured reports, and structured data collection. Free format methods are more flexible and adaptable, however analysis and evaluation is difficult; researchers cannot accumulate or compare data without formal definitions.

Systems Engineering Research in Kongsberg, Norway

In this chapter, we report on the research in systems engineering in Kongsberg in the period 2008-2013. In this subsection, we provide background to the specific situation in Kongsberg: the industrial background, education in systems engineering, and the research model based on master projects. Master projects have been our main research vehicle. The cooperation with industry and the structure of the master program facilitated this form of research.

The Kongsberg Industrial Cluster as Laboratory

The Kongsberg industry inherited a strong technology base from the Kongsberg Våpenfabrikk (weapon factory) in the second half of last century. After privatization in the eighties, the Våpenfabrikk went into a Norwegian version of Chapter 11 for bankruptcy regulations. The stronger parts became strong players in different markets: component production in among others automotive and aerospace, gas turbines, sub sea oil and gas equipment, maritime, and defense. This industry develops and applies complex systems, requiring integral disciplines such as project management and systems engineering. Typical characteristic for most industry in Kongsberg is that the systems operate in harsh environments. For example, sub sea oil and gas equipment operating at depth of 3000m at extreme pressures and temperature ranges, aerospace components and subsystems with lifetimes of 15 years

The Norwegian government, the Kongsberg community and industry and Buskerud University College (BUC) identified Systems Engineering as crucial competence for Norway in 2001. As a result, of national and local government and local industry created the Norwegian Center of Expertise (NCE) for Systems Engineering (SE) as joint effort. They established funding for a 10-year program; see [13].

Master in Systems Engineering at Buskerud University College

One of the main projects of NCE was the establishment of a master study in systems engineering at BUC. The author has described this process in [14]. The goal of the master program is to create an educational path that helps industry to foster new system engineers in 5 to 10 years, rather than the 10 to 20 years in current practice. The result is a master study Systems Engineering that BUC offers in two variants:

Local industry deploys *Industry Master (IM) Students* part-time allowing the students to study part-time. *IM students* need 3 years of elapsed time to obtain a masters title with 2 years of nominal study effort. Typically, *IM students* do not have previous working experience. Since we believe that experience plays a crucial role in developing the systems engineering competence we have introduced this part-time industry involvement. During the study, reflection is encouraged so that students benefit the most from the practical experience in industry.

Part-time students are already working in industry and do their study part-time. Typically, part-time students need at least 4 years of elapsed time to finish their master education.

Master Project Research

All master students have to perform a master project in the last semester of their study. This master project is 30 ECTS[3] or half a year full-time study load. We use the master projects for multiple goals (see [15]) at the same time:

1. To provide value to the industrial employer by working on actual projects,
2. To facilitate students to apply SE in realistic industrial conditions
3. To facilitate the students to make the step from "just applying" to "critical reflection"

---

[3] ECTS stands for European Credit Transfer and Accumulation System and is a standard across the European Union, see http://www.mastersportal.eu/articles/388/all-you-need-to-know-about-the-european-credit-system-ects.html

4. To verify that students are capable to operate at academic level
5. To create a research portfolio and capability at BUC

Master students have done a major part of the research at BUC until now, during their half-year (full-time) master project. Gradually, we start to recruit PhD and other researchers. The results of this research are promising; we have published thirteen papers at conferences and journals (see http://www.gaudisite.nl/MasterProjectPapers.html).

Figure 1 shows the logical order of steps to define a research project in systems engineering. The starting point is a need for improvement in the field, triggered by an industrial problem. Researchers reformulate the problem into an industrial goal; "where do we want to go" instead of "where are we". The problem triggers multiple research questions, e.g. how much can method x mitigate the problem? The next step is to sharpen the research by looking for quantifiable propositions, e.g. full requirements traceability will reduce the change request rate after the project definition with 80%. Researchers can use the research questions and the quantified propositions to formulate a hypothesis as basis for evaluation. Often a number of explicit criteria help in such evaluation. Preferably, the research leaves room to study multiple options. The researcher will need some baseline to evaluate. The baseline can be past performance, benchmarking with other projects or organizations, or comparison of different solutions.

We simplified the research model shown in Figure 1 left for research done in master projects of six months duration. The researcher looks in the systems engineering body of knowledge for methods or techniques that could mitigate the industrial problem. The master student captures the expected improvement in a claim, similar to the quantified proposition in the generic approach. The next step is that the student identifies observables that can support or invalidate the claim. Students can then evaluate the claim by analyzing the observables. The claim and observables form the core of the master project research. The claim relates to the contribution of the method or technique from the body of knowledge. The observables focus on supporting or invalidating this part of the body of knowledge.

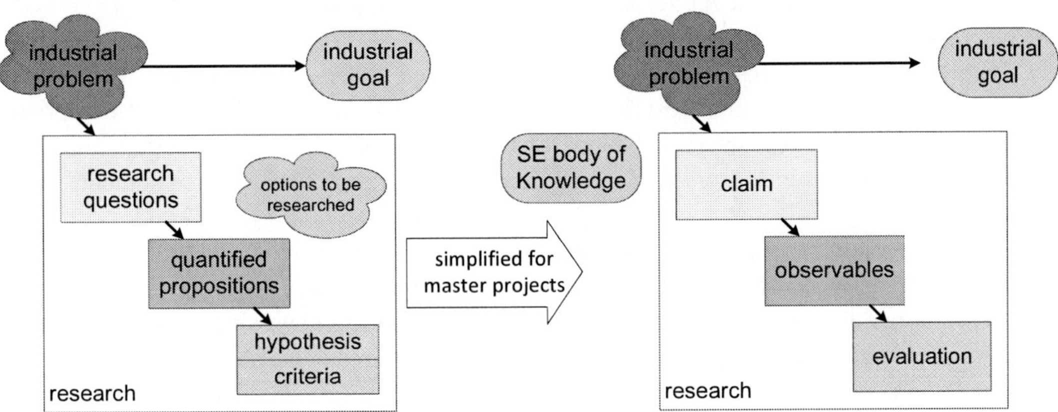

Figure 1. From industrial problem to validated research; left the generic approach, right a simplified approach used for master projects.

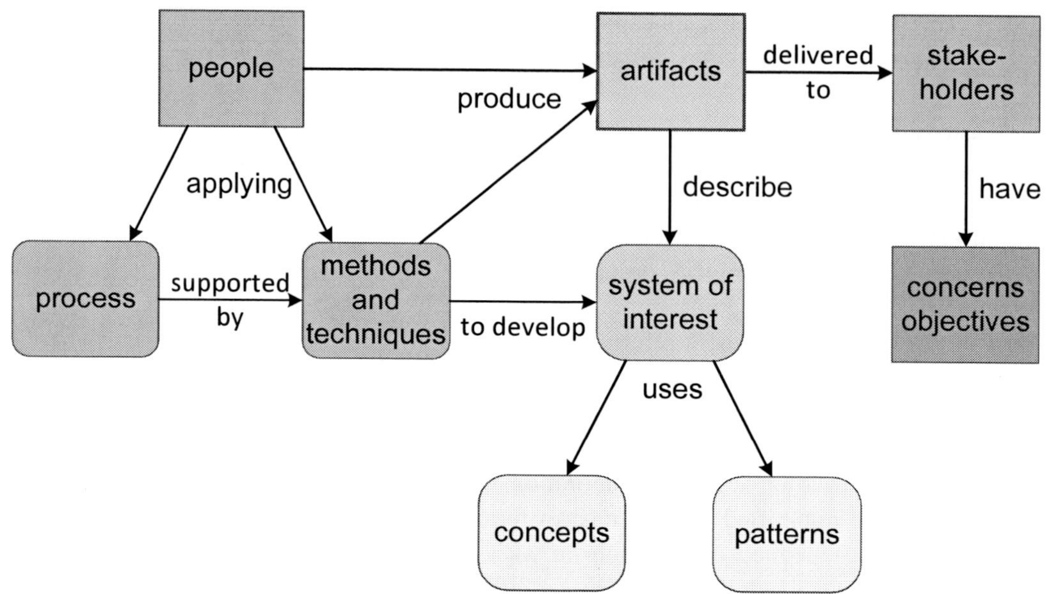

Figure 2. Context diagram with entities relevant for the research.

RESEARCH RESULTS 2008-2013

Local industry and BUC formulated a research agenda in 2008. Core in this agenda is the dominating need for robustness and reliability in harsh environments, a common property of most systems from Kongsberg industry. At the same time, industry feels a continuous pressure to respond to changes and to stay competitive. Reliability and robustness have an inherent tension with the need for change or innovation. The general areas of customer understanding and system design facilitate the research field. Some more specific research fields emerged, such as concept selection, and modeling and analysis. Later we have added the even more generic facilitating field of processes.

Figure 3 shows all research projects in the period 2008-2013 mapped on the Venn diagram representation of the research agenda. The circular dots have been brief exploration projects in the various industries to explore what specific challenges the industrial companies have. The academic supervisors have used this insight to supervise master students in the following years.

For these projects, we have counted the key words to provide an impression of the popularity per subjects. Table 1 shows the highest scoring key words. In the next paragraph we will elaborate the research that relates to the highest scoring key words, since these topics have been researched multiple times, allowing us to reflect on any trends.

Synthesis of Current Research Results

A3 overviews are based on the work by Daniel Borches [16]. He proposes to use A3 size paper (297x420 mm) to document (partial) architecture overviews. The main idea is that the essence of a topic fits on an A3 and that stakeholders can digest this amount of information in a reasonable amount of time. Borches provides a cookbook that recommends combining at least the following views on one A3: partitioning (or physical model), behavior (or functional model), quantification (or Non Functional Requirements (NFR), or key performance parameters), visual aids, and linking information between these views.

Researchers, who started with the cookbook, typically customized the cookbook to fit their problem. For example, Wiulsrød [17] followed the cookbook closely. However, he composed the A3 stepwise while communicating with his stakeholders. Lindtjørn [18] applied it on a manufacturing production line, using more text and less visualization. Frøvold [19] and Kruse [20] used A3s for validation. Validation requires close interaction with stakeholders, especially in the operational domain. Consequence was that visualization has to be close to that stakeholder world; the visualization was less abstract and more realistic, e.g. using photos.

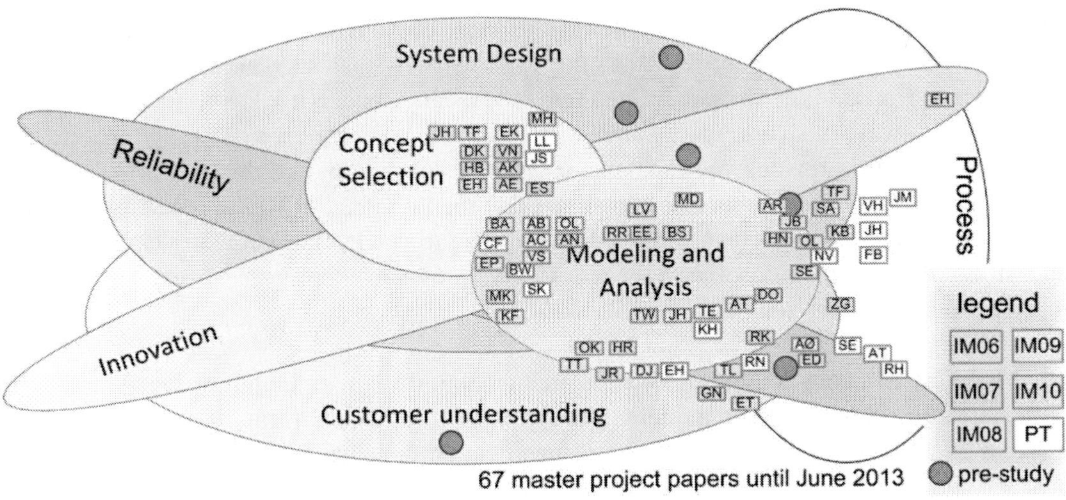

Figure 3. Positioning of research projects on the research agenda.

Table 1. Highest scoring key words for the master project papers in the period 2008-2013

A3 overview	12
Pugh matrix	10
modeling	9
concept selection	9
requirements management	8
verification	5
validation	4
LEAN	4
interface management	4
integration	4
functional modeling	4

Vickram [21] built on the A3 experiences of the previous cohorts. However, he combined it with tooling from Model-Based Systems Engineering (MBSE) creating so-called dynamic A3s. The tools generate dynamic A3s from information in the MBSE repository. These tools allow for linking information; users can navigate multiple A3s via hyperlinks. At the same time, users always see the current baseline.

Pugh matrix and concept selection are closely related, since the primary application of Pugh matrices is concept selection. A Pugh matrix [22] is a matrix to evaluate multiple concepts to select an appropriate concept. However, some researchers applied Pugh matrices slightly different, e.g. for component selection. Some researchers applied alternative ways of selecting concepts. However, most papers in these categories apply Pugh matrix for concept selection. In [23], we have captured our initial experiences based on four case studies. The observation in this paper was that the topic Pugh matrix and concept selection is especially popular in the sub sea oil and gas industry. We speculate that the project pressure in this industry, which invites developers to select the most obvious concept immediately, causes this behavior. This paper describes that the experience is that stakeholders appreciate the Pugh matrix. However, at the same time there is resistance, since it takes time, and it differs from current way of working. Lønmo reports in a recent paper [24] "However, the participants in the previous research papers were reluctant to use the matrix in concept selections. On the contrary, this research shows that the Pugh matrix was appreciated amongst the engineers. All the participants wanted to continue to use evaluation matrices, and nearly all will recommend them to other colleagues." She provides some possible explanations for this more positive reception.

Modeling is a core activity in systems architecting, design, and engineering. The term is used here more broadly than modeling in MBSE. MBSE uses formalized models in for instance SysML to replace mostly verbose specification documents. However, models in the broader sense as used here, are used for understanding, communication, reasoning, and decision making too. Students researching modeling studied aspects such as visualization and effectiveness of the models for understanding, communication, reasoning, and decision making.

Engebakken [25] introduced impact factors to analyze effectiveness of models and to provide guidelines for others when making models. Examples of such impact factors are: close-to-reality, instant recognition, personal relevance, multi-view models, participation, assumptions, iterations, effort and purpose, level of detail, guidance, number of stakeholders, intention, and dynamics. Rypdal [26] and Polanscak [27] build further on the work by Engebakken. Rypdal analyzes how the distance of the stakeholders relates to the kind of model. Polanscak combines modeling with A3 overviews. Stalsberg [28] continues this work relating these impact factors to their role in the phase of the project. The general findings of these four researchers is that modeling is an effective means for understanding, communication, reasoning, and decision making. However, model makers need guidance to make effective models. Lindtjørn [18] reported that effectiveness depends on the skills of the modeler as well. More verbal oriented engineers have difficulties in making effective visual models.

Requirements management is a hot topic throughout Kongsberg industry. There is a clear difference in maturity in systems engineering in general and requirements engineering specifically between companies and business units in companies. The traditional systems engineering domain of defense and aerospace is mature and uses tools like DOORS for requirements management. Sub sea oil and gas, however, is a different world. In this industry standards and regulations from governments, standardization bodies, and oil and gas operators are stacked. Unfortunately, these standards violate requirements management wisdom, such as freedom of implementation, ambiguity, clarity (e.g. the use "may" or "should" rather than "shall"), and scope of requirements.

Tårneby [29] tries to unravel the input requirements in the stack of standards and regulations by implementing a requirements register as spreadsheet. He relates single specific requirements to work packages. Problem is that, even for only one subsea system, the register grows beyond manageability. Tranøy [30] analyzes cost overruns and late variation orders and finds that the tendering and execution process and timing in the oil and gas industry block a proper needs analysis. Consequence is that the incoming requirements do not cover specific operational needs. Hole [31] has similar findings after analyzing non-conformance costs. He proposes a better connection between field study, tendering, and execution in in form of people and documentation.

Hetlevik [32] looks for ways to mature requirements management further in the defense domain by capturing requirements in a more reusable way.

Nysæter [33] and Grinderud [34] apply different tools and formalisms to manage requirements in subsea. The research projects looking into **interface management** also typically look into tools and formalisms. It is too early to draw conclusions on formalisms and tools, other than that MBSE looks attractive for both requirements management and interface management.

Integration, verification, and **validation** are relevant in any project business and development of complex systems. The topics are related since typically both verification and validation takes place during systems integration. Tørnlycke [35] and Henden [36] analyzed problems found in the systems integration tests and traced backward to find causes and potential mitigations for future projects. They found that many problems that arise during systems integration could have been detected in earlier test phases. They also conclude that insufficient documentation of specification and design in earlier phases is often the root cause of integration problems. This in itself is no surprise; main question is whether and how these problems could have been detected and mitigated in earlier project phases. Next question is what the costs are in terms of effort, time, and money, and how these relate to benefits.

Øvergaard [37] and Dalby [38] worked on automated testing to reduce testing effort and to increase testing quality. Research question is whether the effort to create and maintain automated testing is in balance with the benefits. Both papers show that automated testing is beneficial. However, in both cases, the project shows that automated testing is far from trivial. Automated testing mostly makes sense if the testing frequency is increased. Neither project could substantiate the benefits sufficiently, since this requires a significant period of using the automated tests.

LEAN is in fact a collection of LEAN techniques and principles, such as late decision making, set-based design, and A3 reports. Despite popularity of LEAN product development, the introduction in day-to-day practice takes more effort than expected. An example is the study by Hansen [39] in the pre-amble of introducing set-based design in her company. She analyzed the opinions and expectations of the immediate stakeholders and compared them with literature. One of her conclusions is "They [the employees] seemingly need clearer guidance, and ask for access to and knowledge about appropriate tools to use in the implementation. There seems to be a great potential for successful implementation provided the employees gain a better understanding of the method."

Functional modeling is a first essential step to understand and explore systems more conceptually. Teachers are well able to convince students of the relevance of functional modeling. Unfortunately, many engineers in practice work intensely in the physical paradigm, close to the realization, without an explicit understanding of system behavior, e.g. in terms of functions. Several students have tried to introduce functional modeling, for instance in the form of IDEF0 diagrams; see for example the paper by Drotninghaug [40]. A bottleneck for functional modeling seems to be that most functional models are quite abstract, distant from the mental world of many stakeholders. She found that "For the employees without Systems Engineering background the tools that provided the most value were the geographical model, the Context Diagram and the Physical Architecture.", which were the models that are the least abstract and closest to their day-to-day experience.

EVALUATION OF INDUSTRY AS LABORATORY

We have applied industry as laboratory five years in the master projects in Kongsberg. The result is that we have 67 master project papers describing cases where the researchers have applied systems engineering methods or techniques in practice. Thirteen of these papers have been published so far. Of the 67 papers, 18 have been finished in June 2013; we hope to publish four more papers of these recent 18 papers. This means that we publish about 25% of the papers.

We continuously experience the need for a good master project preparation, with proper scoping and identification of a carrier project (e.g. an ongoing project in the company that is willing to accommodate the master project as integral part of the project) and identification of relevant systems engineering methods or techniques that master students can apply for evaluation. We observed some pitfalls and problems:

- Companies that create special projects for master students. Such projects do not fulfill the industry as laboratory paradigm, since the student is more or less isolated from ongoing projects. We estimate that about 10% of the projects are too far outside the main stream.
- Students that select improper systems engineering techniques or methods, e.g. methods that that are too far outside the normal way of working or where the effort is not in balance with expected benefits. We estimate that another 10% of the projects suffer from this problem.

- Dynamics in the company that disturb the chosen carrier or the work of the student. Students regularly suffer from project dynamics. However, in most cases, the dynamics is part of normal practice; good methods and techniques ought to be robust for changes in the context. Students suffer some more from changes in the project timing, which sometimes prevents them to observe expected benefits, since the master project itself has a hard (academic) deadline. Consequence is that the evaluation gets weaker than planned.
- Students that strive too much for a positive outcome. With their engineer's hat on, they need to strive for a positive contribution. However, as researcher they need the opposite drive. Main risk is that students are blind for negative inputs, followed by the risk that the outcome is biased. The academic supervisor has to intervene timely.
- Students and companies that shift the project too far in the process or tool direction. Processes and tools are core assets in systems engineering. However, in the industry as laboratory approach, we favor the application and evaluation of processes or tools in practice, rather than process or tool studies or introduction without application. The application is essential to get some substance in the evaluation.
- Projects so confidential that publication of the results is out of the question. This happens in the defense industry. It also happens in the supplier industry, where the customer does not permit any publication. Sometimes, the researcher can separate and publish the method or technique results without the domain context; loss of context may hamper further analysis of these results.
- Students with high performal capabilities and less linguistic capabilities that have problems in transforming their insights in linear text.
- Well executed projects that are difficult to publish due to a narrow project scope. The findings and data collection may still be relevant for further analysis when the limited scope is taken into account.

The collection of case studies is reaching a size where we gradually hope to harvest slightly broader findings, for example in topics such as concept selection and A3 architectural overviews. Such aggregation would benefit from standardization of data collection. We have to find a proper balance in minimizing constraints and standardization to facilitate aggregation. The first step in this direction has been a workshop and reader on research methods in the preparation phase of the master projects.

CONCLUSION

The master students in systems engineering at BUC who are working part-time in industry are producing a collection of cases of the application of systems engineering in practice. This collection is a first step to substantiate the actual applicability and value of methods and techniques that we have captured in the systems engineering body of knowledge and that we teach worldwide in systems engineering curricula.

The preparation phase and the problems statements formulated in many papers show abundant needs for improvements in systems development, where systems engineering claims to offer solutions. The needs vary over the companies and their systems engineering maturity.

We see the full spectrum of needs from customer understanding and needs elicitation to systems integration, verification, and validation, followed by lifecycle management issues. The practice of systems development struggles with aspects where systems engineering should bring value.

However, main conclusion from these initial case studies is that value and applicability is not self-evident. In general, stakeholders appreciate the application of systems engineering methods and techniques; however, the reception is not that positive that the use of them spreads itself further into the companies.

The research model using master students that have been working in the company and let them practice action research or industry as laboratory has produced a significant number of case studies. With 25% of the case studies actually published, we are creating a collection of case studies to facilitate further theory forming of system engineering

FUTURE RESEARCH

The number of case studies has to increase an order of magnitude to facilitate better substantiation of applicability and analysis of the value of systems engineering methods and techniques. The balance between free format, allowing unbiased observations, and standardization, facilitating aggregation and analysis across cases, needs to evolve, striving for a sweet spot in both aspects. Further theory forming on effectiveness may help in balancing these aspects.

REFERENCES

[1]   Pyster, A., Olwell, D., Hutchison, N., Enck, S., Anthony, J., Henry, D. & Squires, A. (eds). (2012). Guide to the Systems Engineering Body of Knowledge (SEBoK) version 1.0.Hoboken, NJ: The Trustees of the Stevens Institute of Technology ©2012. Available at: www.sebokwiki.org.

[2]   Dixit, I. & Valerdi, R. (2012). *Challenges in the Development of Systems Engineering as a Profession, to be submitted to Journal of Systems Engineering*, preprint 2012.

[3]   A. Squires, A. Pyster, Olwell, D., Few, S. & Gelosh, D. (2009). *Advance Systems Engineering.*" INSIGHT *12* (4), 69–70.

[4]   Boehm B., Valerdi, R. & Honour, E. (2008). The ROI of Systems Engineering: Some Quantitative Results for Software Intensive Systems, *Journal of Systems engineering*, Vol. *11*, 221-234.

[5]   O'Brien, R. (2001). Um exame da abordagem metodológica da pesquisa. In Roberto Richardson (Ed.), Teoria e Prática da Pesquisa Ação [Theory and Practice of Action Research]. João Pessoa, Brazil: Universidade Federal da Paraíba; accessed in the English translation. An Overview of the Methodological Approach of Action Research, University of Toronto http://www.web.ca/robrien/papers/arfinal.html#_edn2, accessed July 25, 2012.

[6]   Potts, C. (1993). Software-engingeering research revisited. *IEEE Software*, Vol. *10*, No.5:19–28, September/October 1993.

[7]   Muller, G. & Heemels, W. P. M., *Five Years of Multi-Disciplinary Academic and Industrial Research: Lessons Learned*; CSER 2007 in Hoboken NJ.

[8]   Bhattacherjee, A. (2012). Social Science Research: Principles, Methods, and Practices, USF Open Access Textbooks Collection, 2012, http://scholarcommons.usf.edu/cgi/viewcontent.cgi?article=1002&context=oa_textbooks, retrieved 2 December 2012

[9]   Glaser, B. G. & Strauss, A. L. (1967). The discovery of grounded theory: Strategies for qualitative research. US New Jersey: Transaction Publishers.

[10]   Strauss, A. L. & Corbin, J. (1994). *Grounded theory methodology: An overview in Handbook of qualitative research*. London: Sage Publications.

[11]   Charmaz, K. (2006). *Constructing grounded theory: A practical guide through qualitative analysis*. Los Angeles: Sage Publications.

[12]   Muller, G. (2013*). Systems Engineering Research Methods;* CSER 2013 in Atlanta

[13]   Bjørnson, T. & Gauteplass, K. (editors). (2013). *NCE Systems Engineering*, http://ekstranett.innovasjonnorge.no/NCE_fs/Infoark%20Kongsberg_engelsk310107.pdf accessed July 5, 2013.

[14]   Muller, G. (2009). *Creating an Academic Systems Engineering Group; Bootstrapping Systems Engineering Research*; CSER 2009 in Loughborough

[15]   Muller, G. (2012). Validation of Systems Engineering Methods and Techniques in Industry; CSER 2012 in St Louis.

[16]   Borches, D. (2010). *A3 Architecture Overviews: A Tool for Effective Communication in Product Evolution*. PhD Thesis, Enschede, The Netherlands: University of Twente.

[17]   Wiulsrød, B. & Muller, G. (2012). Architecting Diesel Engine Control System using A3 Architecture Overview, proceedings of INCOSE 2012 in Rome.

[18]   Lindtjørn, O. C. (2012). *A3 Architecture Overview: A Case Study in a Manufacturing Production Line*, unpublished master project paper at Buskerud University College.

[19]   Frøvold, K. (2011). Applying A3 reports for early validation and optimization of stakeholder communication in development projects, unpublished master project paper at Buskerud University College.

[20]   Kruse, M. (2011). *A3 reports for early validation of new functionality: A case study in the development of maritime operator stations*, unpublished master project paper at Buskerud University College.

[21]   Singh, V. & Muller, G. (2013). *Knowledge Capture, Cross Boundary Communication and Early Validation with Dynamic A3 Architectures*, proceedings of INCOSE 2013 in Philadelphia.

[22]   Pugh, S. (1981). *Concept selection: a method that works*. In: Hubka, V. (ed.), Review of design methodology. Proceedings international conference on engineering design, March 1981, Rome. Zürich: Heurista, 1981, blz. 497 – 506.

[23]   Muller, G., Klever D. G., Bjørnsen H. H. & Pennotti, M. (2011). *Researching the application of Pugh Matrix in the sub-sea equipment industry*, proceedings of CSER 2011 in Los Angeles.

[24]   Lønmo, L. (2013). Concept Selection - Applying Pugh Matrices in the Subsea Processing Domain, unpublished master project paper at Buskerud University College.

[25]   Engebakken, E., Muller, G. & Pennotti, M. (2010). *Supporting the system architect: Model–assisted communication, Systems Research Forum*, Vol. *4*, No.2, 2010, 173-188

[26] Rypdal, R., Muller, G. & Pennotti, M. (2012). *Developing the Modeling Recommendation Matrix: Model-Assisted Communication at Volvo Aero*, proceedings of INCOSE 2012 in Rome.

[27] Polanscak, E. (2011). *Supporting Product Development: A3-assisted Communication and Documentation*, unpublished master project paper at Buskerud University College, 2011.

[28] Stalsberg, B. (2013). *Increasing the value of model-assisted communication: Modeling for understanding, exploration and verification in production line design projects*, unpublished master project paper at Buskerud University College.

[29] Tårneby, T. (2011). *Use of a Register to Classify Customer Requirements for a Subsea Production Control System,* unpublished master project paper at Buskerud University College.

[30] Tranøy, E. (2013). *Reduction of Late Design Changes Through Early Phase Need Analysis*, unpublished master project paper at Buskerud University College.

[31] Hole, J. (2012). *Can a new documentation structure improve requirements management and consequently reduce price of non-conformance to requirements in FMC?* Unpublished master project paper at Buskerud University College.

[32] Hetlev ik, E. (2012). A new Requirement Definition System for Reuse of Requirements, unpublished master project paper at Buskerud University College.

[33] Nysæter, H. (2013). *A requirement management and traceability tool to increase requirements visibility, traceability and reuse in subsea systems*, unpublished master project paper at Buskerud University College.

[34] Grinderud, Z. (2013). *An Analysis of Requirements Management using Siemens Teamcenter: A case study at FMC Technologies,* unpublished master project paper at Buskerud University College.

[35] Tørnlycke, A. (2013). *Visualization of System Integration and Verification by Validation discrepancies using the V-model*, unpublished master project paper at Buskerud University College.

[36] Henden, R. (2013). *Visualization of systems integration discrepancies using the V-model*, unpublished master project paper at Buskerud University College.

[37] Øvergaard, A. & Muller G. (2013). *System Verification by Automatic Testing*, proceedings of INCOSE 2013.

[38] Dalby, E. (2013). *Industry Evaluation of an Automated SW Test Framework Implemented at Unit level,* unpublished master project paper at Buskerud University College.

[39] Hansen, E. Å. (2012). *Set-based design – the lean tool that eludes us; Pitfalls in implementing set-based design in Kongsberg Automotive*, proceedings of INCOSE 2012 in Rome.

[40] Drotninghaug, M., Muller G. & Pennotti, M. (2010). *The value of Systems Engineering Tools for the Understanding and Optimization of the Flow and Storage of Finished Products in a Manganese Production Facility,* proceedings of EUSEC 2010 in Stockholm.

In: Advances in Systems Engineering Research
Editors: Elena Fermi and Adam Lamberti

ISBN: 978-1-62948-310-8
© 2013 Nova Science Publishers, Inc.

Chapter 6

REMOTE AUTOMATIC CHECKING SYSTEM FOR MULTIPLE EMERGENCY LIGHTS VIA INTERNET

Hsiung-Cheng Lin

Department of Electronic Engineering,
National Chin-Yi University of Technology, Taichung, Taiwan

ABSTRACT

Emergency lighting facilities are widely installed in public areas such as school, hospital, station, etc.

According the fire regulation in most of countries around the world, such facilities must be installed and checked regularly to ensure their normal working condition. However, usually they are suspended from a high place and distributed widely in the building.

Presently, the checking responsibility depends on human operation only, taking at least 30 mins for each device. This chapter develops a remote automatic checking system for a number of emergency lights via Internet simultaneously. The proposed system can check the status of emergency lights including the battery, charger and light using a Microprocessor-based Detector in a pre-scheduled period. Every emergency light is assigned a respective identification number.

Then, each Microprocessor-based Detector can transmit the checked results to the Nearby Computer via RS485/232. Using TCP/IP, the Remote Computer can communicate with the Nearby Computer for a long distance via Internet. Accordingly, all checked outcomes received by the Nearby Computer can be transferred to the Remote Computer for data monitoring and recording. Furthermore, the Remote Computer can send the control signals to the Microprocessor-based Detector for a remote real-time checking operation.

Experimental results confirm that the proposed system design presents a good performance in term of robust, stable, and fast response.

I. INTRODUCTION

Nowadays, a variety of emergency lights are widely installed in public buildings for safety requirement in case of fire alarm or temporary electricity failure, continuing to supply lighting electricity for a certain period, e.g. 30 mins. According the fire regulation in most countries around the world, such devices must be examined and maintained regularly to ensure a good working condition.

For example, the battery of emergency light should be checked and discharged for every 2 or 3 months. However, even in the same building these devices are located in different areas. This problem may pose a great difficulty to fully meet the checking criteria if depending on human only. Unfortunately, the checking responsibility still relies on human operation today.

Therefore, the development of remote automatic checking capability for such emergency lights is an indispensable issue in industry. This study describes the development of remote automatic checking system for multiple emergency lights. The proposed system using the wired RS232/RS485 transmission mechanism can transmit a control signal and retrieve the checked results to/from the Microprocessor-based Detector [1-5]. In the case study for emergency lights checking, the VB-based graphical interface is designed to provide commands to start the checking operation and also to receive real-time information from the emergency lights on line [6-15]. Also, all checked result stored in the database can be retrieved once it is requested.

This chapter is organized as follows. Section II gives a profile of the proposed system structure. Section III describes the system software, including GUI, server, client, and 8051 programming.

Particularly, the communication among these devices is demonstrated and discussed for details. In Section IV, the experimental results are presented to confirm that the proposed scheme is capable of remote on-line monitoring and control performance for emergency lights checking via Internet. Conclusions are given in Section V.

II. PROFILE OF THE SYSTEM STRUCTURE

The proposed system structure shown in Figure 1 is to carry out the wired automatic checking for emergency lights via Internet. The status of emergency lights can be checked using the Microprocessor-based Detector in a scheduled term period. Every emergency light with its own Microprocessor-based Detector has its authorized identification number (ID num.: 0-255).

Therefore, the checked results can be transmitted to the Nearby Computer (server) via RS232/RS485 module up to 1200-meter distance. On the other hand, the Remote Computer (client) can communicate with the Nearby Computer based on TCP/IP. As a result, the operation function can be transferred to the Remote Computer for real-time monitoring and control performance.

The checking side mainly includes Microprocessor-based Detector and emergency light, where the Microprocessor-based Detector consists of 8051, A/D converter (0804), Relay Circuit, and Light Detector. Figure 2 shows more details for the hardware structure.

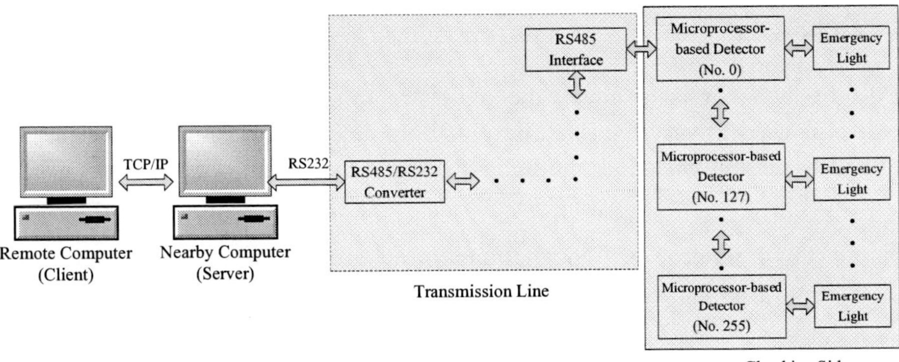

Figure 1. Depiction of the proposed system structure.

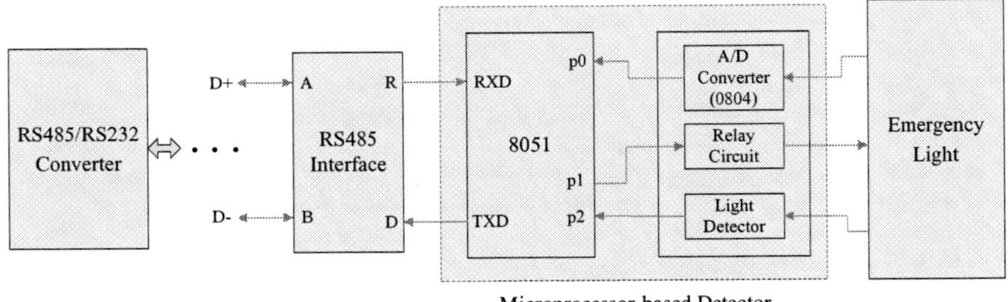

Figure 2. Hardware structure of checking side.

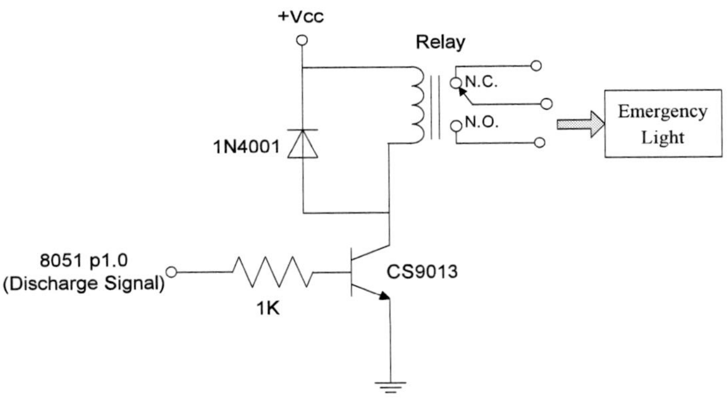

Figure 3. Relay Circuit.

a. Relay Circuit

The Relay Circuit shown in Figure 3 is to receive the command signal from 8051 to discharge the battery for extending its operation life and also checking the light condition. The transistor (CS9013) is used to drive the relay for starting the discharging task.

b. Light Detector

The Light Detector is to check the light strength, shown in Figure 4. The adjusting resistor (50k) is used to vary the sensitivity of light detection. According to equ. (1), will be triggered and turned on whenever the light is on at a good condition, i.e., $V_{cds} \geq 4.11V$. The Q_2, and is thus driven on sequentially. As a result, the resistor will be shorten, and it results in the output $V_H = 1$. In case of light malfunction or at a bad condition, the output is $V_H = 1$, where $V_{cds} < 3.9V$ according to equ. (2).

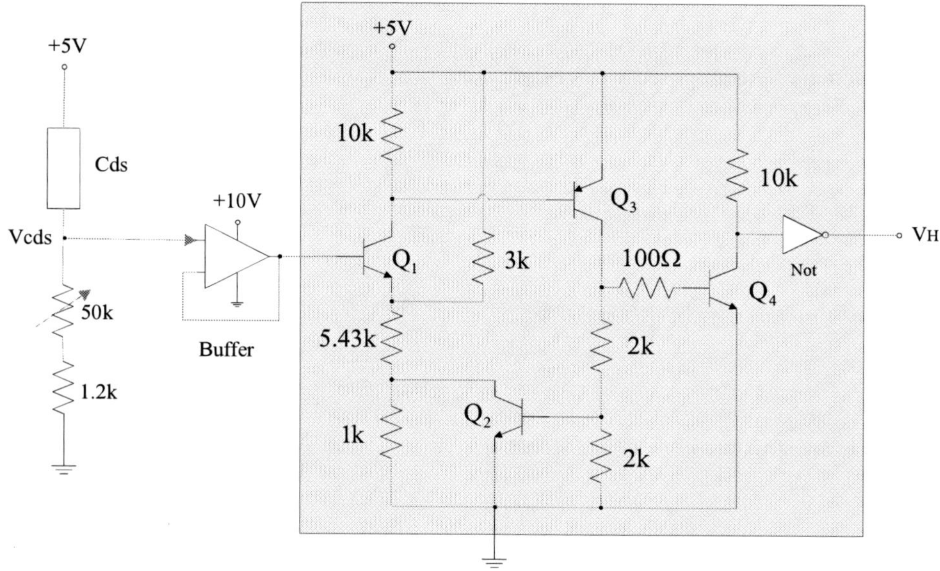

Figure 4. Circuit of Light Detector.

$$V_{Q1H} = \frac{5.43k + 1k}{3k + 5.43k + 1k} \times 5 + 0.7 \cong 4.11(V) \tag{1}$$

$$V_{Q1L} = \frac{5.43k}{3k + 5.43k} \times 5 + 0.7 \cong 3.9(V) \tag{2}$$

III. SYSTEM SOFTWARE

III.1. Graphical Human-Interface

The front panel of nearby computer (server) is designed by VB 6.0, shown in Figure 5. The system allows either manual or automation operation. In automation mode, the checking schedule can be arranged in advance. Once the "System Start" is pressed, the 8051 will be connected with the server for instant checking process. For Web connection, the server will connect with the client by pressing "Start Connection" using TCP/IP.

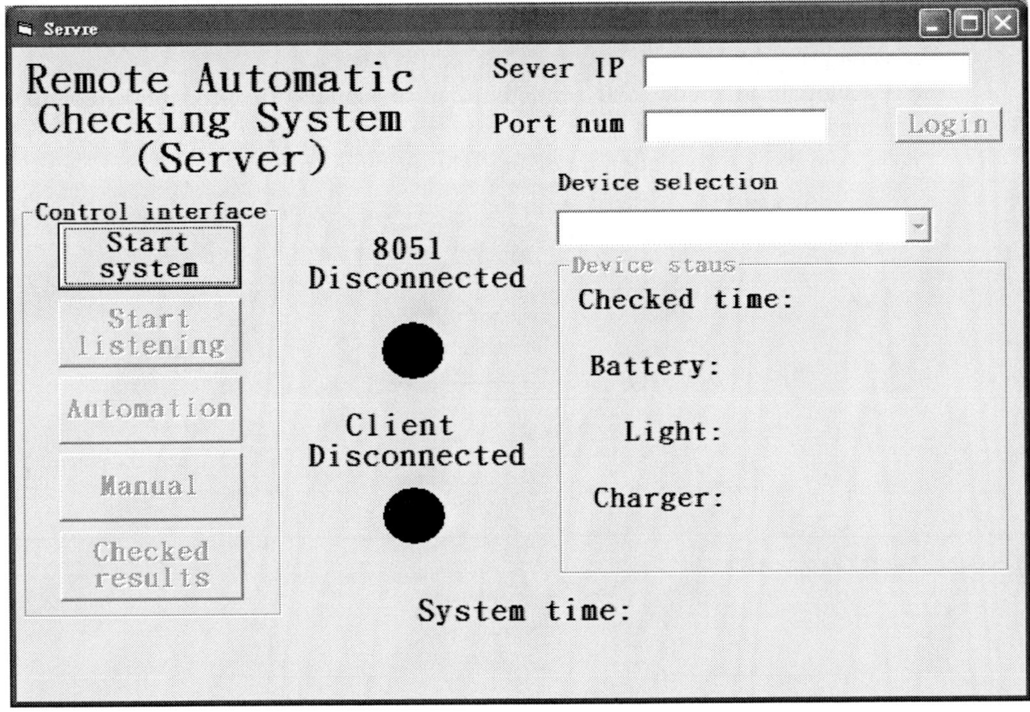

Figure 5. Front panel of Nearby Computer.

III.2. Server Programming

The flowchart of server programming is shown in Figure 6, and described briefly as follows.

(a) Load last setting or re-setting.
(b) Check if the client requests the connection. If yes, send the data to the client. Otherwise, go to next step.
(c) Choose the manual or automation mode.

1. Manual mode:
i.   Check if the manual operation is requested from the client. If yes, go to Step (iii). Otherwise, go to next step.
ii.   Check if the manual operation is activated. If yes, go to next step. Otherwise, go back Step (i).
iii. Send manual commands to 8051.
iv.   Check if any feedback is received from 8051. If yes, go to next step. Otherwise, go back to Step (iii).
v.   Check if the client is connected. If yes, send the data from 8051 to the client. Otherwise, go back to Step (i).

2. Automation mode:

i. Check if the Automation mode is selected. If yes, go to next step. Otherwise, repeat the same procedure.

ii. Start Automation mode, and send the control signals to 8051 following up the scheduled time.

iii. The system is on Automation mode until the system is turned off.

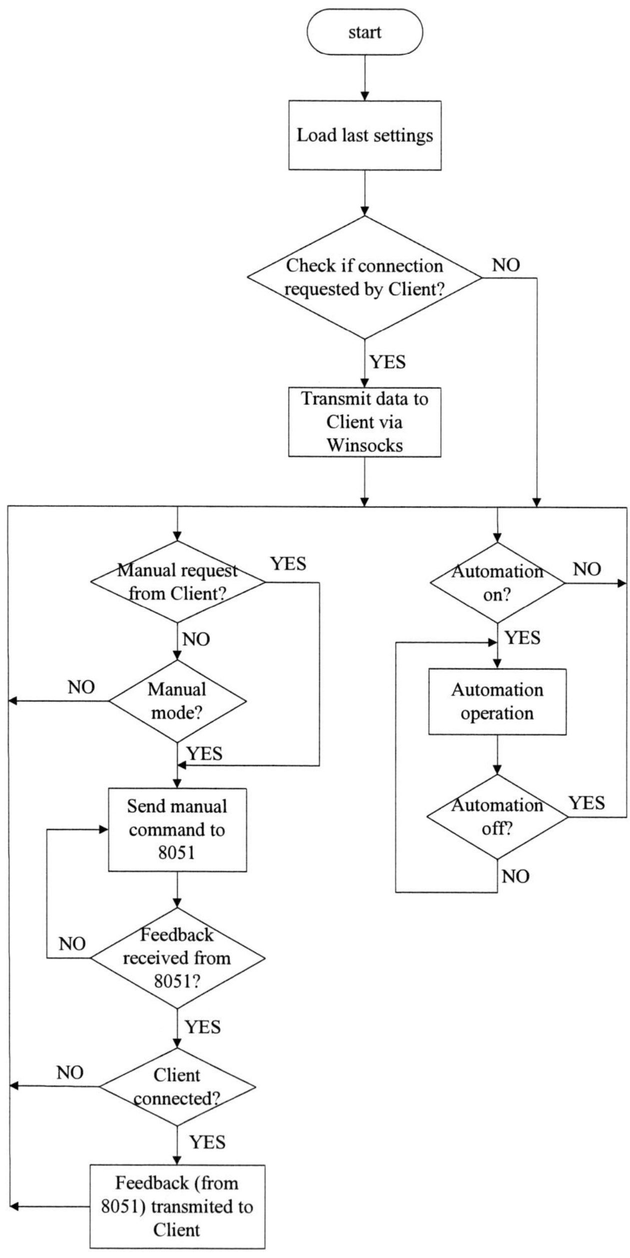

Figure 6. Flowchart of server programming.

III.3. Client Programming

The flowchart of client programming is shown in Figure 7.
Its main procedure is briefly described as follows.

(a) Set IP address of the server for Internet connection, including the channel port.

(b) Check if the connection with the server is requested. If yes, go to next step. Otherwise, go back to Step (a).

(c) Check if the connection is successful. If yes, go to next step. Otherwise, go back to Step (b).

(d) Receive data from the server via Winsocks.

(e) In this step, two operation branches are working in Manual mode and Automation mode independently.

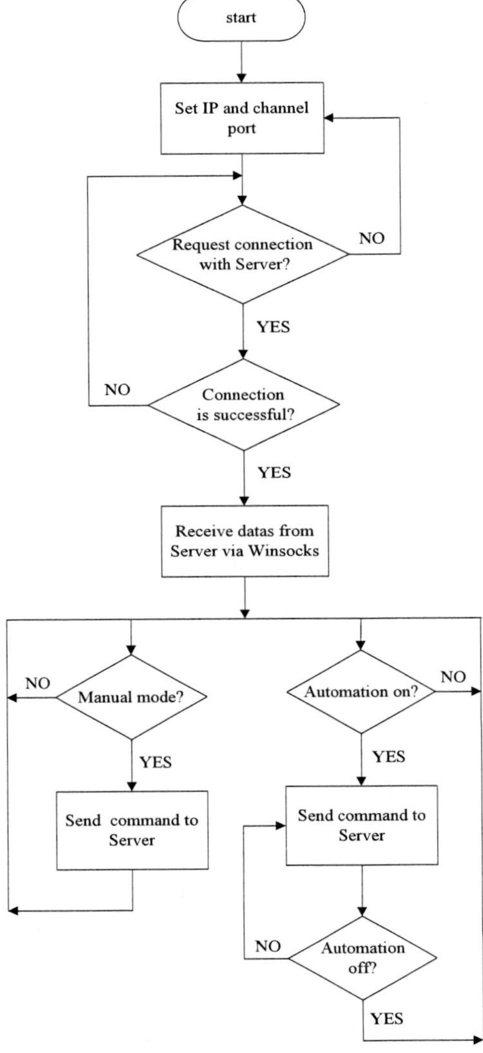

Figure 7. Flowchart of client programming.

III.4. 8051 Communication Programming of Microprocessor- Based Detector

The programming flowchart of microprocessor (8051) in the Microprocessor-based Detector is shown as Figure 8.

(a)   Set COM1 as the serial communication with the outside world. SCON is set to as transmission mode 1, and TMOD is set as timer mode 2. For details, see Table I.

Table I. Register setting status

SCON	TMOD	SMOD	TH1	Bps	TI	RI
50H	20H	1	FDH	9600	0	0

(b)   Continue the Step (b) until the connection request from the server is received and then go to next step.
(c)   The data (amount of both feed and water) is read to the p2 of 8051 from A/D converter, and it is stored temporarily in the serial port buffer (SBUF) ready for transmission to the server.
(d)   Go back to Step (c) until upon complete transmission of data from UART serial port (TXD) to VB (server), and TI will be set to 1.
(e)   Set time delay.
(f)   Read the SBUF data (feeding amount and time schedule) received from the VB.
(g)   Go back to Step (f) until upon complete data receiving, and RI will be set to 1.
(h)   Operate the command.
(i)   Go back to step(c) until the system is turned off.

III.5. Checking Programming

When the checking side receives the command, the Microprocessor-based Detector will soon start the checking performance. The flowchart of emergency light checking is shown in Figure 9. The procedure is briefly described as follows. Note that the charger provides 4.8~5.3V to charge the battery (3.6V 700mAh).

(a)   Check if the checking signal is received. If yes, go to next step. Otherwise, repeat the same procedure.
(b)   With the A/D converter, acquire the battery voltage.
(c)   Check if the battery voltage is 4.0~4.7V. If yes, the system of emergency light is confirmed malfunction. Otherwise, go to next step.
(d)   If the battery voltage is 4.8~5.3V, it confirms that charger OK, battery bad and light unknown. If the battery voltage is less than 3.4V, it confirms that charger bad, battery unknown, and light unknown. Then, go to Step (i). If the battery voltage is 3.4~3.9V, go to next step.
(e)   Check if the light is bright. If yes, go to next step. Otherwise, it confirms that charger OK, battery unknown, and light bad. Then, go to Step (i).
(f)   Discharge the battery for 30 mins.

(g)   Check if the light is bright. If yes, it confirms that charger OK, battery OK, and light OK. Then, go to Step (i). Otherwise, it confirms that charger OK, battery bad, and light OK. Then, go to Step (i).

(h)   Transmit the checked results to the Nearby Computer/ Remote Computer.

(i)   Check if stop the system. If yes, the system stops. Otherwise, go back to Step (a).

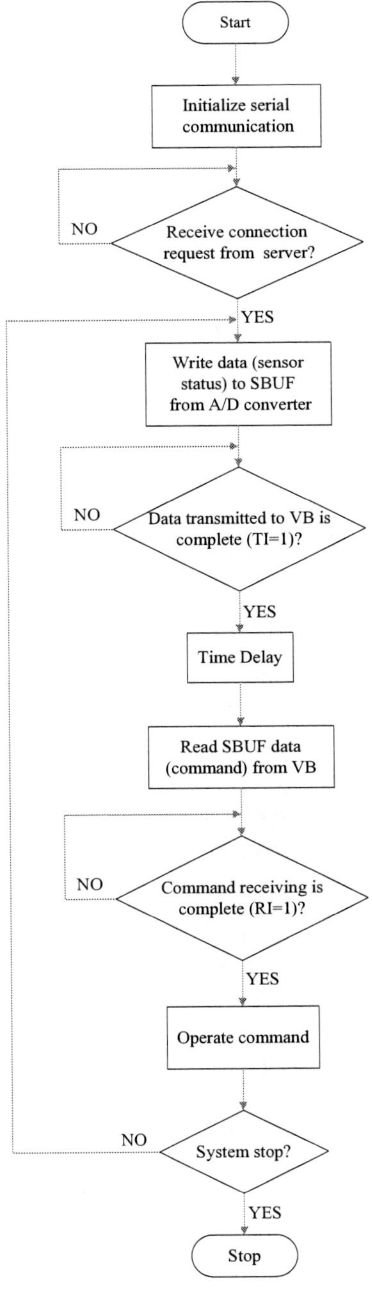

Figure 8. Flowchart of 8051 programming.

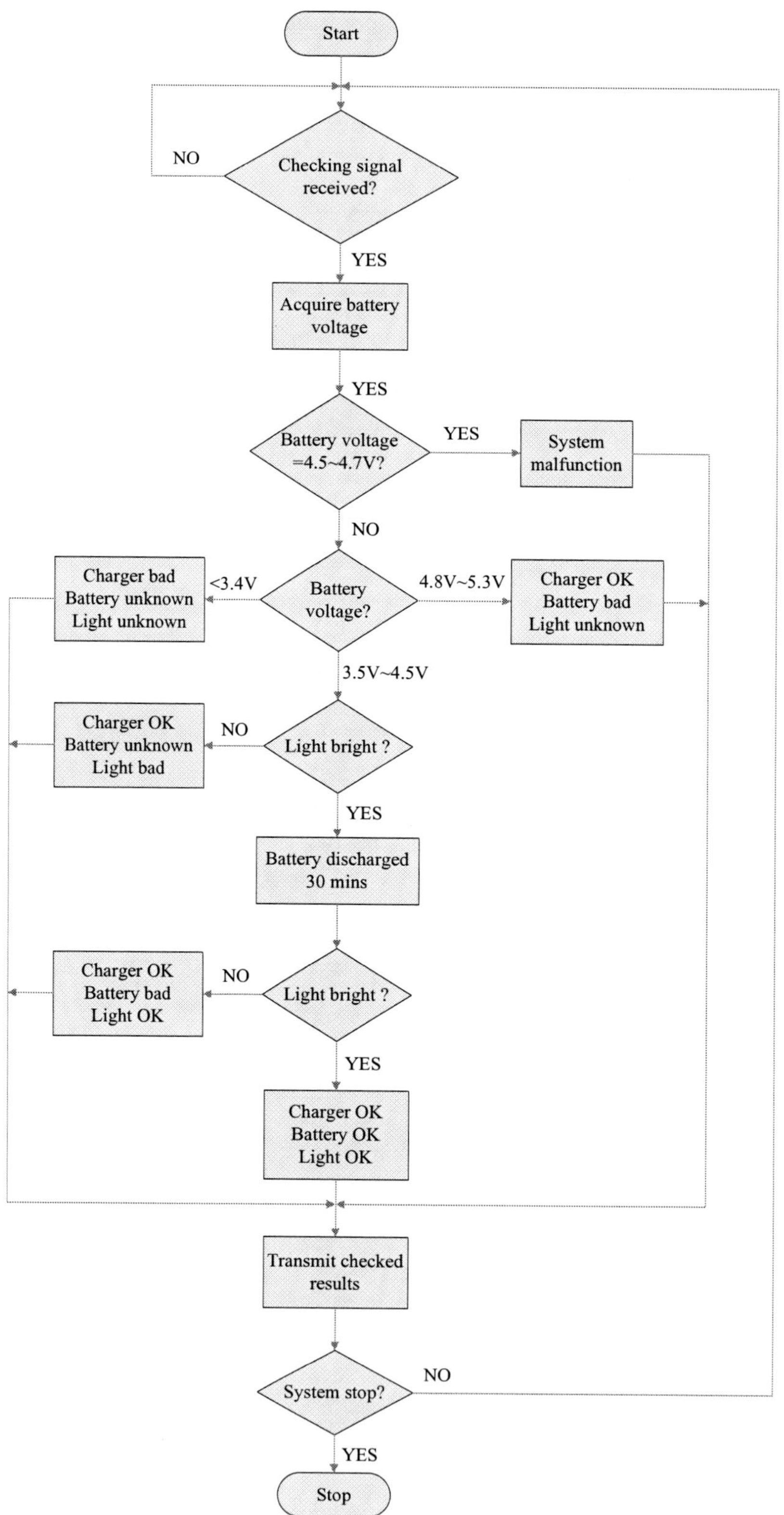

Figure 9. Flowchart of emergency light checking.

Based on Figure 9, if the system of emergency light is found malfunction, its status is unknown.

More details about the relation between the checked results and representative codes are concluded in Table II.

Table II. Codes of checking results with status

Codes	System	Charger	Battery	Light
0000	Malfunction	x	x	x
0001	Normal	Normal	Disorder	x
0010	Normal	Disorder	x	x
0011	Normal	Normal	x	Disorder
0100	Normal	Normal	Low power	Normal
0101	Normal	Normal	Normal	Normal

Note: x means unknown status.

IV. EXPERIMENTAL RESULTS

The proposed system can carry out a remote checking using either automatic or manual operation for multiple emergency lights via Internet.

Therefore, the client from a remote location can view and even control the checking process.

All checked results can be also saved in the database for further investigation. The following procedures have indicated its performance details.

- Connect sever with 8051

The first step to implement the proposed system is to connect server with 8051 via RS232.

Once the "Start System" button, shown in Figure 10, is pressed, the connection will be operated immediately whenever the 8051 is standby.

- Connect client with server

The real-time operation of the proposed system in the both client and server is shown in Figure 11. Figure 11(a) shows the system status before connection. Right side indicates that 8051 is connected.

Left side indicates client is not connected with server. Figure 11(b) shows that 8051, client and server are working together on line.

Figure 10. Startup of system connection.

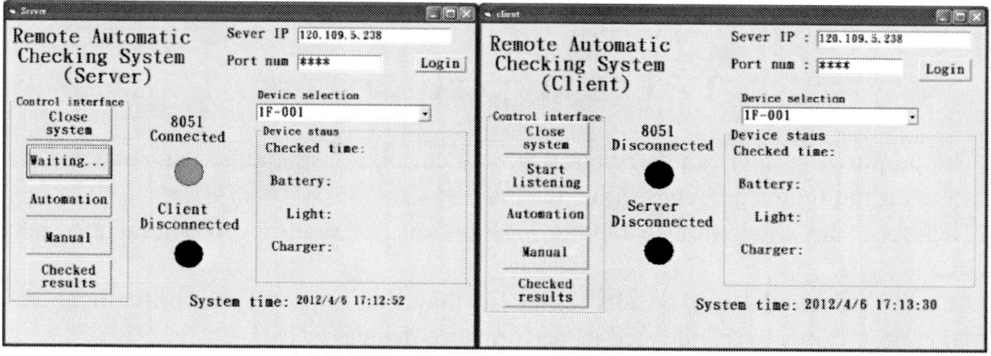

(a) Right: client status, left: server status

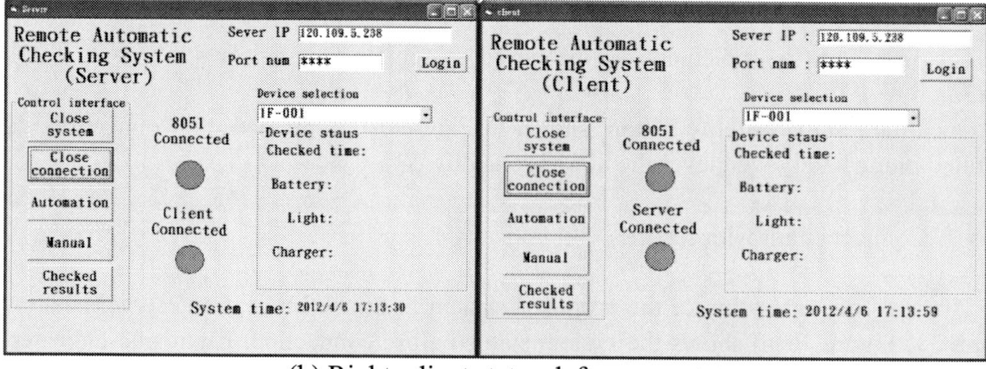

(b) Right: client status, left: server status

Figure 11. Connection between client and server (a) before connection (b) after connection.

- Web checking from client

With Web connection, the client can carry out a remote monitoring and control function for emergency light checking. Figure 12 shows the checked results based on automation operation. The manual operation is shown in Figure 13.

(i) Automation operation

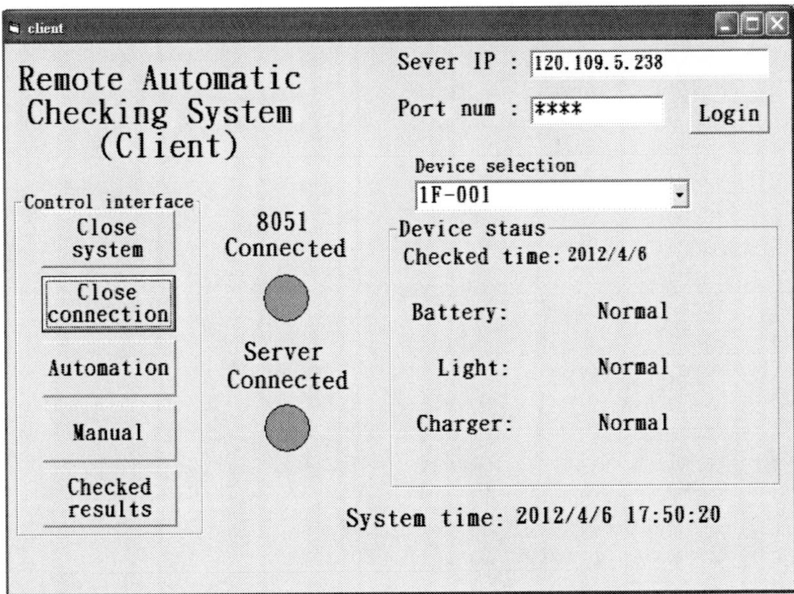

Figure 12. Automation operation in client.

(ii) Manual operation

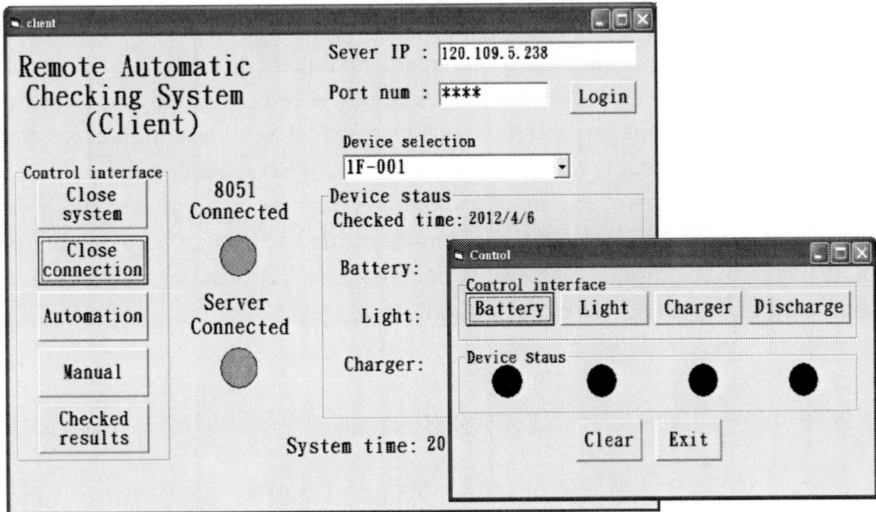

Figure 13. Manual operation in client.

- Checked results

The checked results obtained from the Microprocessor-based Detector can be transmitted to both client and server. It can be retrieved from the database, shown in Table III.

Table III. Checked results

Emergency lighting facilities

Device	Battery	Light	Charger	Date	Remarks
1F-001	Normal	Normal	Normal	2012/4/6　17:50:20	
1F-002	Normal	Normal	Normal	2012/4/6　17:50:20	
1F-003	Low power	Normal	Normal	2012/4/6　17:50:20	
1F-004	X	Normal	Normal	2012/4/6　17:50:20	
1F-005	Normal	Normal	Normal	2012/4/6　17:50:20	
2F-001	Normal	X	Normal	2012/4/6　17:50:20	
2F-002	Normal	Normal	X	2012/4/6　17:50:20	
2F-003	Normal	X	X	2012/4/6　17:50:20	
2F-004	Normal	Normal	Normal	2012/4/6　17:50:20	
2F-005	Normal	X	Normal	2012/4/6　17:50:20	
3F-001	Normal	Normal	X	2012/4/6　17:50:20	
3F-002	Low power	Normal	Normal	2012/4/6　17:50:20	
3F-003	Normal	Normal	Normal	2012/4/6　17:50:20	
3F-004	Normal	Disorder	Normal	2012/4/6　17:50:20	
3F-005	Normal	Normal	X	2012/4/6　17:50:20	
4F-001	X	Normal	Normal	2012/4/6　17:50:20	
4F-002	Low power	Normal	Normal	2012/4/6　17:50:20	

記錄: I◄ ◄ 31 之 31 ► ►I ► ✕ 無篩選條件　搜尋

CONCLUSION

Although emergency lights must meet the fire regulation in most countries around the world, it is still a very difficult task to be realized due to heavily depending on human operation today. Therefore, this chapter has developed a remote automatic checking system for multiple emergency lights successfully. The graphical interface authorizes the users to operate the checking system on line from a remote computer. Accordingly, the e-platform checking system allows the client to carry out a real-time monitoring and even control process. Based on RS485 transmission, the communication distance between the emergency lights and the Nearby Computer can be extended up to 1200 meters, avoiding a possible interfere occurred from such as wireless methods.

REFERENCES

[1]　Chao Xue, Xiang Cheng, Yang Guo, Yong Lia, "The design of a sub-nanojoule asynchronous 8051 with interface to external commercial memory", *IEEE 8th International Conference on ASIC* (ASICON '09), pp. 427-430, 2009.

[2]   Rech, P., Gerardin, S., Paccagnella, A., Bernardi, P., Grosso, M., Sonza Reorda, M., Appello, D., "Evaluating Alpha-induced soft errors in embedded microprocessors", *15th IEEE International On-Line Testing Symposium,* pp. 69-74, 2009.

[3]   Rech, P., Paccagnella, A., Grosso, M., Sonza Reorda, M., Melchiori, F., Loparco, D., Appello, D., "Evaluating the Impact of DfM Library Optimizations on Alpha-induced SEU Sensitivity in a Microprocessor Core", *IEEE Transactions on Nuclear Science,* Vol. 57 , Issue 4 , Part: 1, pp. 2098-2105, 2010.

[4]   Leite, F., Balen, T., Herve, M., Lubaszewski, M., Wirth, G., "Using Bulk Built-In Current Sensors and recomputing techniques to mitigate transient faults in microprocessors", *10th Latin American Test Workshop*, pp. 1-6, 2009.

[5]   C.H. Chen, H.C. Lin, Y.C. Liu, W.C. Hsu, "Local-loop based robot action control module using independent microprocessors", *Computer Applications in Engineering Education* (CAE), Vol. 18, Issue 4, pp. 593–606, 2010.

[6]   H.C. Lin, "On Yaoxiang Li, Zhiping Wang, "A VB-Based Forest Field Data Collection System", Second International *Workshop on Computer Science and Engineering*, Vol. 2, pp. 551-554, 2009.

[7]   line Internet-based Multi-location Power System Harmonics Analysis and Monitoring", *Computer Applications in Engineering Education* (CAE), Vol. 17, Issue 3, pp. 241-252, 2009.

[8]   H.C. Lin, "Web-based Remote On line Maximum Wind Power Monitoring and Control System", *Computer Applications in Engineering Education* (CAE), Vol. 15, Issue 2, pp.155-165, 2007.

[9]   H.C. Lin, L.Y. Liu, K.H. Pai, "A Microprocessor-Based Monitoring and Control System via Internet in Case Study of Pet Care", Advanced Materials Research, *Computational Materials Science*, Vols.268-270, pp. 772-780, 2011.

[10]   C.H. Chen, H.C. Lin, "A Distance e-Learning Platform for Signal Analysis and Measurement using FFT", *Computer Applications in Engineering Education* (CAE), Vol. 19, Issue 1, pp.71–80, 2011.

[11]   C.H. Chen, H.C. Lin, Y.C. Liu, W.C. Hsu, "Local-loop based robot action control module using independent microprocessors", *Computer Applications in Engineering Education* (CAE), Vol. 18, Issue 4, pp. 593–606, 2010.

[12]   H.C. Lin, "On line Internet-based Multi-location Power System Harmonics Analysis and Monitoring", *Computer Applications in Engineering Education* (CAE), Vol. 17, Issue 3, pp. 241-252, 2009.

[13]   H.C. Lin, "Web-based Remote On line Maximum Wind Power Monitoring and Control System", *Computer Applications in Engineering Education* (CAE), Vol. 15, Issue 2 , pp.155-165, 2007.

[14]   H.C. Lin, "An Internet based graphical programming tool for teaching power system harmonics measurement", IEEE *Transactions on Education*, Vol. 49, No. 3, pp.404-414, 2006.

[15]   H.C. Lin, "An interactive framework for power system harmonics measurement using graphical programming and the Internet", *Computer Applications in Engineering Education* (CAE), Vol. 14, Issue 1, pp.44-52, 2006.

INDEX

D

E

T

U

V

W